Welcome to:

An innovative approach to EMP (electromagnetic pulse) protection that can allow entire communities to maintain their "pre-pulse" lifestyle.

Many things could cause the long term loss of the electric power grid upon which we all depend, including geomagnetic storms and cyber-terrorism.

High-altitude EMP is a unique form of destruction, however, that adds the loss of most of our vital electronic infrastructure to the electric power grid devastation. The electronic infrastructure includes telecommunications of all kinds, inventory control, personal computers and most forms of modern transportation.

In this book, you will learn how to use electromagnetically shielded solar photovoltaic panels to provide electricity through all kinds of electric power grid outages, whether the outage lasts for hours or for years.

You will also learn relatively inexpensive means for electromagnetically shielding volumes from the size of a pantry to a large building in order to protect the items inside from EMP.   New methods of financing EMP protection for communities -- from homeowners associations to county governments -- are also addressed.

Don White, PE, MSEE
EMP Solutions & Renewable Energy Creations, LLC

Jerry Emanuelson, BSEE
Futurescience, LLC

Summary of this Handbook

This book is the first volume of a planned five-volume series on EMP protection.

Although the Internet has many articles on EMP history as well as sources providing survival foods, water purification, general medications, etc., all of these things do little to maintain one's lifestyle after an EMP event. Many people knowledgeable about EMP protection are frustrated that the government is essentially doing nothing to protect the civilian sector from the EMP threat.

The mission of this book is to act as a catalyst, providing a few templates from which small communities and small governmental units, such as towns and counties, can begin specific programs to provide EMP protection of homes and commercial buildings and also provide protected solar rooftop backup power.

Individual owners of homes and businesses can benefit from the information in this book, but the information is mainly directly toward promoting a community-level of EMP protection and resiliency of the local electrical and electronic infrastructure.

THE GOALS ARE:

Phase 1 -- To initiate several pilot projects of EMP protected villages and towns in locations across the United States over a two-year span.

Phase 2 -- To EMP protect most villages to small cities over the following eight years using all of the information learned in Phase 1.

The extra bonus is that Phase 2 will generate many new EMP and solar related products and services, opening new markets and generating seven million jobs in the process.

Catastrophes such as severe geomagnetic storms resulting from coronal mass ejections from the sun could collapse the power grid, leaving large sections of this critical infrastructure out of service for years. Cyber-attacks could also cause a lengthy outage of large chunks of the power grid. The EMP protection plan explained in this book isolates communities from much of the adverse effects of these lesser catastrophes as well.

This handbook is written for the general public with additional sections to provide information for relevant professionals. It is illustrated by 105 images composed of drawings, graphs, diagrams, charts, tables, spreadsheets, and photos to support the text and tutorials. Most of the contents have not been published before.

Disclaimer

Estimates and projections contained herein are prepared by EMP Solutions and Renewable Energy Creations, LLC (REC) and are based on information currently available (May, 2013) and that developed by REC. The latter involves many visits and discussions, many engineering and economic calculations, subjective judgments and analysis. While believed sound, no representation or assurance is made as to their accuracy or attainability.

The author tried to contact engineers and others in the EMC (electromagnetic compatibility) community. Almost none of these professionals provides information contained in this handbook. It is believed the reason may be that EMP hardening details are classified because of restrictions placed on Department of Defense and other related activities performed by EMC contractors. This book has been written without the benefit of any kind of access to classified information.

Further, recognize that the protection of a building, home, or solar-PV installation is a function of many variables. Thus, it is nearly impossible to generalize performance and predict future outcome. While several product photos are shown, courtesy of a named company or other source, this does not mean that that product is manufactured for EMP application. Rather, it is representative of the product, with some modification, that may be used for EMP protection of homes, buildings and solar installations. Thus, it is concluded that this book is a protection starter for an EMP attack that is nearly inevitable. As the Boy Scout motto says, "Be Prepared." So, do your own due diligence.

Finally, recognize while severe geomagnetic storms and cyber-attacks on the power grid would be catastrophic beyond anything U.S. residents have ever faced, they fall far short of the impact in comparison to the results of an effective nuclear EMP attack.

About 50% of the 100 images (illustrations) used in this book are developed by the author, especially almost all tables, spreadsheets, drawings, diagrams and graphs. However, some photos pose a problem. For example, Tineye, an Internet inverse image search engine site allows the user to search for sources of photos from their 2.1 billion inventory. On the Swiss chalet with solar rooftop, for example, they produced 106 hits – making it impossible to identify the real and original source for credit. Thus, some source photos are unknown and cannot be credited. The publisher acknowledges that the Cover image is copyright, by the Commonwealth of Australia."

Printed in the United States of America
Library of Congress PCN: (in process)
Library of Congress Cataloging-in-Publication Date:
Library of Congress Card Number: (in process)
ISBN-13: 978-1484909850
Sku:

Acknowledgments from the Publisher

This handbook, **EMP – Protect Family, Homes and Community** is based on scientific and engineering principles. This handbook and treatise are dedicated to the Glory of God – the Epitome of Truth – a word incapable of practice by man.

Even humble Diogenes of Sinope, the Greek philosopher who walked throughout Athens carrying a lantern in daylight, searched for, but could not find, an honest man.

While trying our best, this book falls short of providing much known/unknown truth in an environment often shrouded by claims, representations, distortions and intentional omissions. Yet, we try, our creativity notwithstanding.

The publisher expresses thanks to those who reviewed this handbook including George Bingley, Wayne Pilikian, Bill Duff, Peter Prior, Joan Piikian and Don White's wife, Colleen.

The publisher will contribute 10% of all book sales to charity including *Grace Presbyterian Church* of Lake Suzy, FL 34269, USA, the *Wounded Warrior* Project seen on TV, and SonLight Power, a small solar installation company who, for free, brings both solar electricity and the Bible to schools, clinics and churches in impoverished locations in Central America and Africa.

Enjoy.

EMP Solutions
www.emp-safeguard.com

Contents

Contents, Continued

Figures and Images

Figures and Images, Continued

8

This page is presented to suggest what life may be like in a post EMP incident without EMP hardening to show, by contrast, the benefits of protection described in this book.

One expected scenario for a town in USA after an EMP incident at mid afternoon, in a world of no EMP civil building & no solar-PV protection.

Day 1, after EMP

School was letting out when students noticed almost all vehicles were frozen-like in different street positions. Their cell phones and ipads were dysfunctional to call 911 or home. Nothing electronic seemed to work. One student said this looks like "One Second After", a book his dad was reading. He said a high altitude nuclear event would wipe out electronics and electrical for hundreds of miles.

Students had to walk home where they noticed electricity was out; houses were dark and TV and radio were dead. Neighbors gathered discussing the perplexing and scary events. One observed EMP-like atmosphere with only a few older vehicles moving.

Day 2

A large crowd gathered at the town gazebo as the mayor spoke. He said he would talk loudly as the public address wasn't working and the back up generator didn't work. He said not to use the hospital as they have no functional lights, generator or any kind of power. He spoke of a death at a grocery store as patrons fought over remaining canned food on the shelves. Some stole food. Medications were running out at pharmacies.

Day 7

A few had already died as the dialysis and oxygen machines didn't function. Bottled water was gone. Toilets would not flush and toilet paper was gone. No running water as there is no water tower. Most people began to smell. Food and drink shortage caused some house burglaries as desperate folks turned primitive. A few were shot.

Day 30

Scores of premature deaths from water and food starvation and from food poisoning. No undertaker or funeral services. Decedents were buried in large pits in nearby woods. Most horses and cattle and all pigs were gone. Discussions turned to killing pets for food. House and store burglaries were routine. Almost no contact with neighboring towns or elsewhere as no telecom and travel vehicles. No airplane, train or bus.

Day 90

Now massive deaths from starvation and water. Salmonella, and other disease became rampant and no medication available anywhere. One helicopter did land from a distant seaport. But, the packages were raided, savage-like from desperation. All pets have been eaten and few fish and birds remained. Talk initiated about eating those of recent death with no disease. Ammunition for self protection was running low. Emergency services became dysfunctional due to no fuel, oxygen, and operable vehicles. Hope was all but gone as there was little expectations of any relief vehicles or prospects of their appearance.

What if this threat had been a Solar Storm instead of an EMP?

This book emphasizes the upper-atmospheric, nuclear electromagnetic pulse (EMP) threat protection vs. the solar flare protection since the former is far more damaging. To protect the electric grid from a GeoMagnetic Storm (GMS), special surge suppression is needed at the point(s) of failure at the power generation, step-up and step-down transformers at substations.

While both EMP and a severe geomagnetic storm, may take out the electric grid in a designated area, a GMS would not immediately disrupt transportation and would result in essentially all electronics and electrical devices still in working condition. If the food and water grocery stores, pharmacies and gasoline stations, have back-up generators, then life is severely curtailed but not as depicted here with an unprotected nuclear EMP incident.

The illustration below presents an overview of the general impact of a nuclear EMP incident vs. a geomagnetic storm, although, unless protected, the geo-magnetic storm can destroy a substation transformer servicing a long trans-mission line. A GMS has a very-low frequency radiation (in other words, a long wavelength, nearly a DC signal, coupling mostly into very long, high-voltage power transmission lines. Since a building may have wires not longer than 500 feet (0.1 miles), a GMS couples into a building with a level of (0.1mi/100mi) = 0.1% of the level of a power line. For a private home with 50 foot (0.01 miles) long wires, a GMS couples into homes and internal devices with less than 0.01mi/100mi = 0.01% of the level of transmission lines. So it follows that a GMS would not damage small portable electric or electronic devices.

· General Identification of EMP and Solar Flare threats and survivors

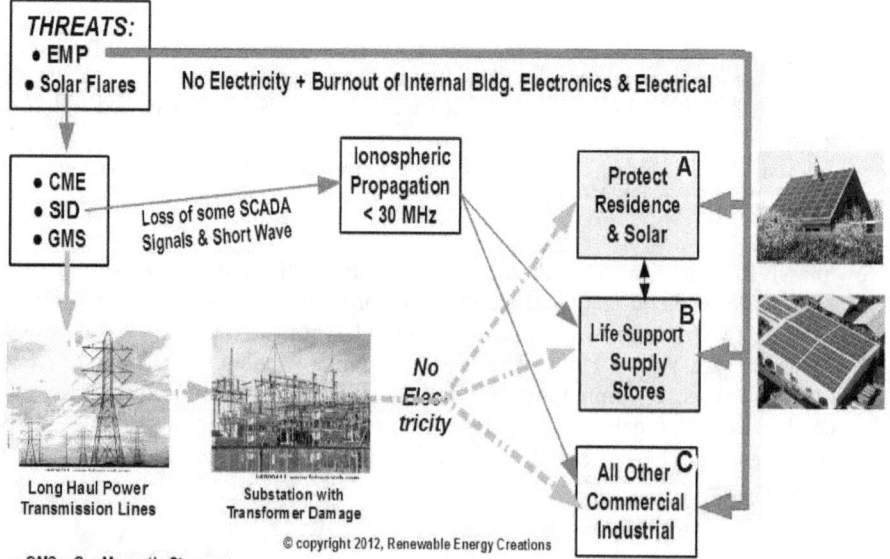

© copyright 2012, Renewable Energy Creations

- GMS = GeoMagnetic Storms
- SID = Sudden Ionispheric Disturbances
- CME = Coronal Mass Ejections

A = residential with full building and solar protection
B = stores carrying food, water medicines and gasoline

Another way of viewing this handbook, is to examine the progress of the unfolding EMP hardening as different components are added. ("hardening" is the military/government vernacular for preventing electric burnout damage). This may be seen in the diagram below. Follow the sequencing:

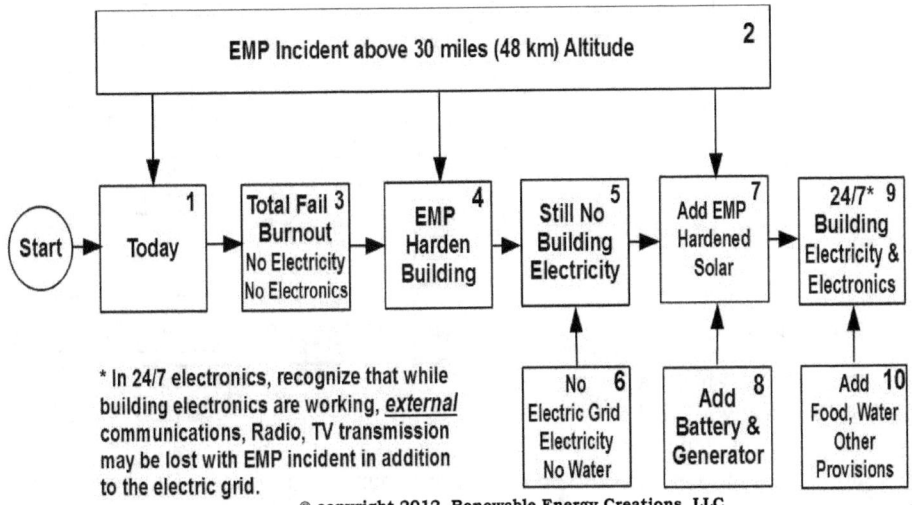

© copyright 2012, Renewable Energy Creations, LLC

Box 1 refers to the status of the global structure today, May 2013, with no civil infrastructure EMP hardening, hoping the event (box 2) will never happen. Most informed people believe the situation is "not if, but when." So, when it occurs, box 3 shows a burnout in subject building (commercial or residential) electronics and most/all damage or burnout of building electrical. Electrical means lighting, heating, air conditioning, hot water, appliances, etc). Many/most of the electrical devices have electronic controls; so, they, too, are lost.

The building is then EMP hardened (shielded, bonded, grounded, surge suppressed and filtered (box 4) as explained in Chap. 5. This protects a building and all its electrical and electronic contents from burnout. But, there is still no electricity (box 5) from the electrical grid (box 6).

Next step is to add a EMP-protected, solar-PV rooftop (box 7). Because there is no active electric grid, and solar is not operational at nighttime and overcast skies, batteries and a generator are added (box 8). These three sources then produce a 24-7 availability of electric power (box 9). It remains to add stored food, water, other provisions to ensure survival since the availability of these sources may be unknown (box 10).

Not shown above are the many enclosures, other than buildings and homes, such as vehicles, aircraft, railroad engines, pleasure boats and ships. The same EMP coupling physics applies along with most of the same solutions.

How to Use this Handbook

Based on their EMP knowledge and topics of interest, the diagram provides recommended general sequence to follow in our readers' priorities. This does not mean that the reader will not want to read other chapters as some chapters depend upon information conveyed in earlier chapters.

Every chapter has a chapter overview of 1/2 to one page in length.

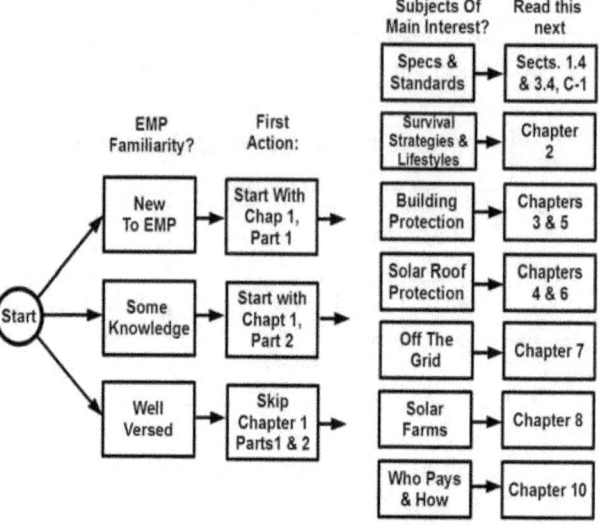

Recommended Sequence of Reading Chapter Material

So, the 10 chapter abstracts can be read in less than one half hour. This constitutes a summary of the entire book, This will help the reader flag out which chapters and sections he may wish to return to for more in-depth reading.

As remarked in a previous section, this handbook contains about 100 images to enhance the presentation and to develop a better understanding The non-technical reader will appreciate this additive over more traditional word discussions only. However, in some cases, the additional information gained will be more appreciated by engineering readers to provide more rationale and comprehension. For those sections, they will be identified by *italics* **at the session number and title** meaning that the lay reader can skip that section as not necessary to further his understanding. This process (*flagging with italics)* will save the non technical reader time by not reading info primarily of interest to the technocrat. Of course, the non-technical reader is always welcomed to read any section if he so chooses.

A few of the images have been repeated more than once in different chapters. The relevance of the image to different applications is the reason. However, a few images may appear to be the same but are actually a step-growth process in further development and expansion of the drawing and topic.

Some chapters have references for recommended sources of additional information. The index, appearing just before the appendices, has been expanded to over 500 entries (a number rarely found in books of this type) to help speed finding information sought. The appendices are expanded to provide supporting material for greater in-depth understanding.

Benefits of this Handbook

There are 15 sectors of principal readers of this book who may financially benefit from this information. There is another score of organizations and professionals also involved. Several million folks in all in USA, plus foreign, specifically:

(1) EMP government and industry planners who have fiduciary responsibility to ensure that their country, in some meaningful way, contributes to a timely EMP awareness, planning and protection.

(2) The Electromagnetic Compatibility (EMC) Community, since they can provide most of the products and services needed for EMP and solar hardening.

(3) Electric utilities because their enormous infrastructure is one principal target of a EMP incident.

(4) Relevant subject regions (in USA, metro cities, municipalities, counties, and their 3,140 economic development offices and their advisers.)

(5) Relevant to #(4) in dissemination, corresponding Chambers of Commerce.

(6) Homeowners Associations to watch dog developments to ensure that their 25 million households receive some degree of timely EMP planning and protection.

(7) Manufacturers of solar panels, inverters and other solar-related items.

(8) Solar distributors and wholesalers, regional and local sales offices.

(9) Builders and solar rooftop installers and maintenance personnel.

(10) Financing sources, investment bankers, brokers and insurance companies.

(11) Commercial and industrial facility building owners.

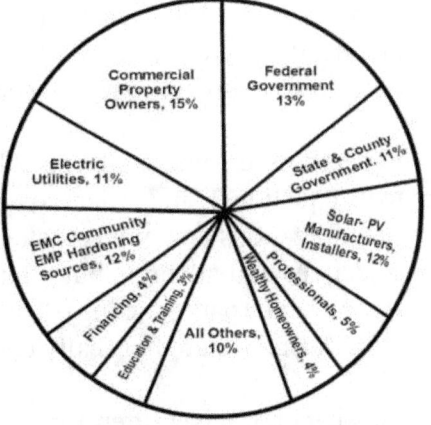

(12) Big chain and franchise hotels/motels and shopping malls since their losses experienced from an EMP incident would be catastrophic.

(13) Education and training centers, such as seminar and short-course providers, universities and even senior high schools and trade schools for helping to become more aware, and participating in the education process.

(14) Professionals: architects, engineers, investment bankers, hedge funds, real estate investors, appraisers, zoning, media, Internet networking, developers, trade associations, expo and conference centers because they will all become involved.

(15) EMP Survivalists, perhaps numbered at between 500,000 and 2,000,000 defined as individuals and families acting as a separate entity for survival (see Chap. 2)

(16) Readers will also be provided, at a low optional cost, software to do prediction and analysis of shielding effectiveness calculations, and break even and ROI for solar installations.

Preface

The author of the original editions of this book, Don White, is a professional engineer by education, training, author, lecturer, and CEO of his own technical companies over a 35-year span. Don is not a politician, snake-oil salesman, turf warfare king, or opinionated, judgmental hypocrite. All of which are disdained with tolerance.

I say the above because unlike the other 13 technical books I have written and published, this one sometimes interfaces with the above type of people. I am not a polished orator, nor a professional persuader, deceiver, and liar. I will often appear to be quite simple, but rely heavily on facts, numbers, truth, and honesty. I try to qualify (sometimes quantify) opinions.

In the fields of the Electromagnetic Pulse and solar energy, I am dealing with a subject different than electromagnetic compatibility (EMC), my home base, for which I am (was) known worldwide. So expect to see me dealing with the multifacets of EMP and solar from different directions, many unknown to some practitioners.

Expect to see some math (no calculus; a little algebra, a touch of trig, a few equations) in order to make reality out of the dangers of generalizations and empty statements. As such, you will see little use of adjectives and adverbs, but a lot of numbers. Over 200 years ago Lord Kelvin said, (paraphrasing) "Unless and until we start speaking in terms of numbers, we are not really communicating."

This handbook is written for the non-technical reader. Later on, another version may be written for (1) the engineer and technical people, and (2) those in banking and finance. However, both types will find this handbook interesting and informative now as I am pioneering and hopefully speaking your language, but with a non-technical emphasis. I am also reminded that pioneers get arrows in the back.

Since I hope this publication will become a reference document for some time, there is some intentional repetition to insure that a user searching for some specific information, will not lose sight of some underlying caveat or point that was either already mentioned or to-be-mentioned elsewhere in the book. For that reason, this book contains about 7% intentionally redundant information.

By the way, please read page 9. It describes what life may be like without EMP protection following a EMP incident. The contrasts are reasons for this handbook. From lessons learned from such scenarios, which could become all too real, we have proposed alternative scenarios for EMP protecting communities while increasing economic expansion and employment.

****

One problem that arose in publishing this book was in how to depict a high-altitude nuclear explosion in diagrams. A daytime high-altitude explosion makes a very brief flash in the sky and nothing else would be visible. The appearance of a nighttime high-altitude nuclear explosion would be very different depending upon a large number of variables.

The problem of how to graphically depict a high-altitude nuclear explosion was solved by the fact that one of the companies that publishes this book uses a nuclear explosion in its logo. The specific image is of the Castle-Romeo nuclear test from a barge in the Pacific Ocean, which, of course, is at sea level.

So please don't write to us telling us that we know nothing about nuclear explosions. This is just a graphic symbol and we really do know better.

****

There are references in this book to **EMP Victims** and **EMP Survivors**. It is important to keep in mind that these victims and survivors refer to electrical and electronic **equipment and devices**, and not to human victims and survivors.

Chapter 1

Part 1 - EMP - A Historical Review and Summary

Chapter 1, Part 1 Overview

This chapter is an in introduction to EMP (Electromagnetic Pulse in general, and HEMP = High Altitude EMP, in particular) and provides an historical review. Early emphasis is on the US Department of Defense studies, tests, engagements, and applicable military standards.

In general, only strategic nuclear forces, certain military communications systems, U.S. Navy ships, certain other military units, a few U.S. government buildings and classified sites are the only U.S. sectors and infrastructures that are EMP hardened today. Therefore, the entire civilian commercial, 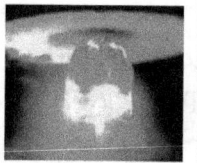 industrial, and residential sectors plus electric utilities, transportation, and most other government buildings are unprotected to a EMP incident and would become victims of an electric grid blackout and burnout.

The electric grid is also susceptible in a very different way to the more intensive low-frequency geomagnetic storms. In contrast to nuclear EMP, geomagnetic storms will have very little to no impact on buildings and on solar rooftops. However, they serve to further underscore the importance and rationale for independent power backup and availability at the building level.

If the you encounter any unfamiliar terminology, you may wish to review Appendix A which contains over 100 definitions of terms used in this handbook. Also, portions of each chapter are marked as being supplementary material for technical people. Non-technical people can skip these sections without missing anything.

1.1 What are EMP and HEMP?

The acronym, EMP, stands for electromagnetic pulse, which can take many different forms. An upper atmospheric (or high altitude EMP, called *HEMP*) burst is the radiation from a nuclear explosion above the earth at altitudes above about 20 miles (32+ km). HEMP couples into cables and all metallic objects within line-of-sight of the explosion.

The physics and results the EMP from a nuclear explosion are very different depending upon the altitude. A low altitude nuclear explosion will cause widespread death to humans, destruction to infrastructure and vehicles. Contrast this to a HEMP burst which will cause no immediate damage to humans (maybe heart pacemakers will burn out) and little to no obvious physical infrastructure damage. Since the acronym *HEMP* is confused with hemp cloth, rope and marijuana in Internet searches, the "H" is dropped in the remainder of this book. So EMP should be understood to be **high altitude** nuclear EMP. Other types of EMP are beyond the scope of this book.

Note: Independent of increasing threat level from all types of nuclear adversaries, electronic based, microprocessor devices dysfunction will increase with (1) the increased use of microelectronics, (2) the lowering of microprocessor power requirements (and, therefore, maximum operating voltage levels), and (3) increases in microprocessor speed and bandwidth. Example: vehicles older than early 1980s use no microprocessors and are far more immune to EMP; whereas, vehicles may become totally dysfunctional in later models.

Electromagnetic radiation from EMP may disrupt or burn out electronic (and some electrical) devices, whether connected or not to the electric power grid. This electromagnetic energy radiation also directly couples to building solar rooftops and to solar farm installations.

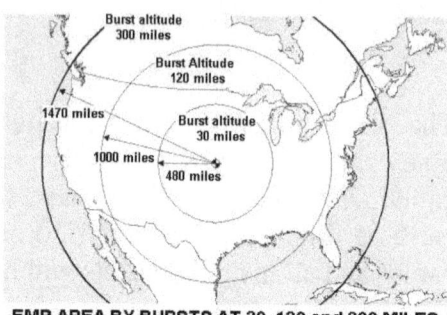

Fig. 1.1 – EMP, an upper atmospheric explosion. (In the daytime, there would be nothing but a brief flash.)

Fig. 1.2 – EMP coverage area at 30, 120 and 300 miles (Gary Smith, EMP Pulse Threats)

One perceived worst-case scenario of a EMP incident is: If a nuclear adversary of any type with a scud-like missile with nuclear warhead and the ability to explode it high above an American city were to do so, it would have a massive effect in all directions. Almost immediately all communications systems in line-of-sight would be disrupted completely (may be burnt out). In that case, there would be no radio, no television, no telephone and no Internet. Indeed, no electricity except certain protected buildings and certain classified government sites. Installations somewhat beyond explosion line-of-sight may be disrupted, but not burned out due to diffraction.

The following paragraph is a *"High altitude EMP"* described by Dr. George W. Ullrich, Deputy Director, Defense Special Weapons Agency:

A nuclear weapon detonated at high altitude releases some of its energy in the form of gamma rays. These gamma rays collide with air molecules and produce what are called Compton electrons. The Compton electrons, in turn, interact with the earth's magnetic field, producing an intense electromagnetic pulse that propagates downward to the earth's surface. The initial gamma rays and resultant EMP move with the speed of light. The effects encompass an area along the line of sight from the detonation to the earth's horizon. Any system within view of the detonation will experience some level of EMP. For example, if a high-yield weapon were to be detonated 400 kilometers (250 miles) above the United States, nearly the entire contiguous 48 states would be within the line-of-sight. The frequency range of the pulse is enormously wide -- from below one hertz to one gigahertz. Peak electric fields can reach tens of thousands of volts per meter. All types of modern electronics are potentially at risk, from Boston to Los Angeles; from Chicago to New Orleans.

Systems and Equipment Impacted and EMP/EMI Hardening
(for technical readers)

Supervisory Control and Data Acquisition (SCADA) systems are electronic control systems that control electrical transmission and distribution, water management and oil and gas pipelines across the United States. Together with digital control systems (DCS) and programmable logic controllers (PLC) they find extensive use in these equipments and systems. The U.S. power industry invests about $1.4 billion annually in new SCADA equipment, 50 times the reinvestment rate in transformers for transmission. Between 25 and 30 percent of the protection and control equipment is upgraded or replaced annually, with each new component generally being more susceptible to EMP damage than its predecessor because of the low-level of energy required for its operation (and a proportionately lower level of energy required for its destruction).

Shield critical SCADA and communications components.

To protect from the effects of EMP one can either acquire EMP hardened electronic equipment, or provide volume shielding inside a room or cabinet for non-hardened electronics. Shielding, bonding, grounding and filtering all can provide protection. For example, EMP hardening a telecommunications cabinet may cost well into thousands of dollars, depending on size.

MIL-STD-188-125-1 provides over 100 pages of detailed information on protecting ground based facilities and equipment from the effects of high altitude EMP. This is discussed later in this chapter.

Electric Power Grid

Three regions exist in the United States that have nominal 60-Hz frequency independence from one another, and one region's collapse would not necessarily lead to the collapse of the others. However, the sub regions, which serve about 70% of the North American population, are for organizational purposes only and do not have frequency independence from each other. A sufficiently large disturbance, like a major EMP event, will likely cause the power grid within a whole region to collapse through cascading failures. Also, as mentioned below, the grid is sensitive to larger geomagnetic storms, which result from certain kinds of intense solar activity.

Electrical power is the product of voltage and current. Electrical resistance loss is proportional to the square of the current. Therefore, it is most efficient to transmit power at the minimum practical current, which means the highest possible voltage. Standard values for modern alternating current (AC) transmission lines range from 115 kV to 765 kV, with currents up to a few thousand amperes.

Distribution to the end users of electricity typically involves voltages ranging from 13.5 kV to 69.5 kV, to be stepped down again to the 120 or 240 volts used in most U.S. households. Distribution has sub-stations, like transmission, only smaller, that are not manned, being run entirely through electronic controls.

Geomagnetic Storm Grid Susceptibility (not an EMP phenomenon)

There are about 2,000 large transformers rated 345 kV or above, about 18,000 generating plants rated 20 MW or above and about 14,000 large (10+ transmission lines each) substations in the United States. Until very recently, no large transformers had been produced in the U.S. for decades. Worldwide capacity for these units, which are built to order and take 12 to 18 months to manufacture, is under 100 per year, with most going to China and India. Including transportation, delivery time for large transformers ordered today is three years. The U.S. replaces about 20 of its large transformers every year. So a loss of even 10% or 20% of the large transformers in the U.S. would take several years of worldwide production to make good, with no assurance that foreign manufacturers would give the U.S. priority over orders for other customers already in the pipeline. The two U.S. plants that have recently started production could not produce those transformers if the power grid were down;

When magnetic fields move around in the vicinity of a conductor, such as wires, and cables, a geomagnetically induced current is produced in the conductor. This happens on a grand scale during geomagnetic storms (the same mechanism also influences long telephone lines and long rail lines and pipelines. Long transmission lines (many kilometers in length) are subject to causing power grid damage by this effect. This chiefly includes operators in China, northern Europe, North America, and Australia, especially in more modern high-voltage, low-resistance lines. The European grid consists mainly of shorter transmission cables, which are less vulnerable to damage.[2,3].

The (nearly direct) currents induced in these lines from geomagnetic storms are harmful to electrical transmission equipment, especially generators and transformers, inducing core saturation, constraining their performance (as well as tripping various safety devices), and causing coils and cores to heat up.

This heat can disable or destroy them, even inducing a chain reaction that can overload and blow transformers throughout a system. [4,5,6] This is what happened on March 13, 1989: in Québec, as well as across parts of the northeastern U.S., the electrical supply was cut off to over 6 million people for 9 hours due to a huge geomagnetic storm. Some areas of Sweden were similarly affected.

According to a study by Metatech corporation, a storm with a strength comparative to that of 1921 would destroy more than 300 transformers and 130 million people would be without power (a cost totaling several trillion dollars.[6] A massive geomagnetic storm could knock out electric power iin large sections of the grid for months.[7] [*This statement should be contrasted with a single EMP incident above 50 miles altitude which, in addition to destroying many transformers in the electrical grid, destroys the electronic contents of millions of buildings and may leave millions of vehicles dysfunctional in their tracks.*] By receiving geomagnetic storm alerts and warnings (for example, by the Space Weather prediction Center; via Space Weather satellites as SOHO or ACE), power companies can minimize damage to power transmission equipment by momentarily disconnecting transformers or by inducing temporary blackouts. Preventative measures also exist, including preventing the inflow of geomagnetically induced currents into the grid through the neutral-to-ground connection.[1]

References #1:

[1] "A Perfect Storm of Planetary Proportions" IEEE *Spectrum*. 2012-02. Retrieved 2012-02-13.

[2} Natuurwetenschap & Techniek Magazine, June 2009

[3] http://192.211.16.13/curricular/ENERGY/0708/articles/solar/SolarForecast07SkyTel.pdf
   Solar Forecast: Storm AHEAD

[4] Severe Space Weather Events: Understanding Societal and Economic Impacts

[5] Metatech Corporation Study

[6] Severe Space Weather Events: Understanding Societal and Economic Impacts : a Workshop
   Report. Wash. D.C.: National Academies, 2008 Web. 15 Nov. 2011. Pages 78, 105, & 106

[7] "Massive solar flare 'could paralyse Earth in 2013'". The Daily Mail. September 21, 2010.

1.2 EMP Historical Overview

(The following material in this section has been extracted from various reliable sources including the web site of Jerry Emanuelson, B.S.E.E, Futurescience, LLC.)

The fact that an electromagnetic pulse is produced by a nuclear explosion was known since the very first days of nuclear weapons testing, but the magnitude of the EMP and the significance of its effects were not realized until later years.

During the first U.S. nuclear test in 1945, electronic equipment was shielded because of the expectation of an electromagnetic pulse from the detonation. All signal lines were completely shielded, in many cases doubly shielded. In spite of this, many records were lost because of spurious radiation pickup at the time of the explosion that paralyzed the recording equipment.

During British nuclear testing in 1952-1953 there were instrumentation failures that were attributed to "radioflash," which was then the British term for EMP. The high altitude nuclear tests of 1962, as described below, increased awareness of HEMP beyond the original small population of nuclear weapons scientists and engineers.

Starfish Prime

On July 1962, a 1.44 megaton USA nuclear test in space, 250 miles (400 kilometers) above the mid-Pacific Ocean, called the Starfish Prime test, demonstrated to nuclear scientists that the magnitude and effects of a high altitude nuclear explosion were much larger than had been previously calculated.

The Thor missile carrying the Starfish Prime warhead actually reached a maximum height of about 680 miles (1,100 km), and the warhead was detonated on its downward trajectory when it had fallen to the programmed altitude of 240 miles (400 km).

Starfish Prime also made EMP effects known to the public by causing electrical damage in Hawaii, about 898 miles (1,445 km) away from the detonation point, knocking out about 300 streetlights, setting off numerous burglar alarms and damaging a telephone company microwave relay link. The EMP damage of the Starfish Prime test was quickly repaired because of the ruggedness (compared to today) of the electrical and electronic infrastructure of Hawaii in 1962. Realization of the potential impacts of high-altitude nuclear EMP became more apparent to some scientists and engineers during the 1970s as more sensitive solid-state electronics began to come into widespread use.

The relatively small magnitude of the Starfish Prime EMP in Hawaii (about 5,600 volts/meter) and the relatively small amount of damage done (for example, only 1 to 3 percent of streetlights extinguished) led some scientists to believe, in the early days of EMP research, that the problem might not be as significant as was later realized.

Newer calculations showed that if the Starfish Prime warhead had been detonated over the northern continental United States, the magnitude of the EMP would have been much larger (22 to 30 kilovolts/meter) because of the greater strength of the Earth's magnetic field over the United States, as well as the different orientation of the Earth's magnetic field at high latitudes. These new calculations, combined with the accelerating reliance on EMP-sensitive microelectronics, heightened awareness that the EMP threat could be a very significant problem.

Little is known about the history of how scientists in the Soviet Union came to an understanding of the unique aspects of EMP There are indications that the Soviet scientists, like the United States scientists, took a long time to completely understand the high-altitude EMP phenomenon. One of the first open scientific publications about nuclear EMP was by the noted Russian scientist Aleksandr S. Kompaneets in 1958. Kompaneets theory was assumed to be accurate by many people who did not have access to the actual classified data until a 1964 report by Victor Gilinsky of the Rand Corporation pointed out a number of significant errors in the Kompaneets analysis.

During the United States Starfish Prime nuclear test, the Soviet Union stationed scientific expeditionary ships in the Pacific near the Johnston Island launch point and at the southern conjugate region (at the opposite end of the geomagnetic field line from Johnston Island) near the Samoan Islands. In addition to general scientific data gathering, the Soviet scientists obviously gathered valuable data in preparation for their Operation K high-altitude tests that were carried out over Kazakhstan three months later.

During one of the Soviet high-altitude tests, the Soviet scientists instrumented a 570-kilometer (350-mile) section of telephone line in the area that they expected to be affected by the nuclear detonation in order to measure the electromagnetic pulse effects. The monitored telephone line was divided into sub-lines of 40 to 80 kilometers (25 to 50 miles) in length, separated by repeaters. Each sub-line was protected by fuses and by gas-filled overvoltage   protectors. The EMP from one of the 1962 nuclear tests caused all of the fuses to blow and all of the overvoltage protectors to fire in all of the sub-lines of the 350-mile monitored telephone line.

Published reports, including a 1998 article in the "IEEE Transactions on Electromagnetic Compatibility" have stated that there were significant problems with ceramic insulators on overhead electrical power lines during the Soviet nuclear tests of 1962. In 2010, a technical report written for a United States government laboratory, Oak Ridge National Laboratory, stated, "Power line insulators were damaged, resulting in a short circuit on the line and some lines detaching from the poles and falling to the ground."

The only nations to detonate nuclear weapons in outer space are the United States and the Soviet Union. The U.S. program began in 1958 with the *Hardtack Teak* and *Hardtack Orange* shots, both 3.8 megatons. These warheads were initially carried on Redstone rockets. Later tests were delivered by Thor missiles or the XM-33 Strypi for Operation Fishbowl tests, and modified Lockheed X-17 missiles for the Argus tests. The purpose of the shots was to determine both feasibility of nuclear weapons as an anti-ballistic missile defense, as well as a means to defeat satellites and manned orbiting vehicles in space.

(The next paragraph and diagram are of technical interest only)

Fig. 1.3 shows how the peak EMP on the ground varies with the weapon yield and burst altitude. Note that the yield here is the prompt gamma ray output measured in kilotons. This varies from 0.1-0.5% of the total weapon yield, depending on weapon design. The 1.4 megaton total yield 1962 Starfish test had an output of 0.1%, hence 1.4 kiloton of prompt gamma rays.

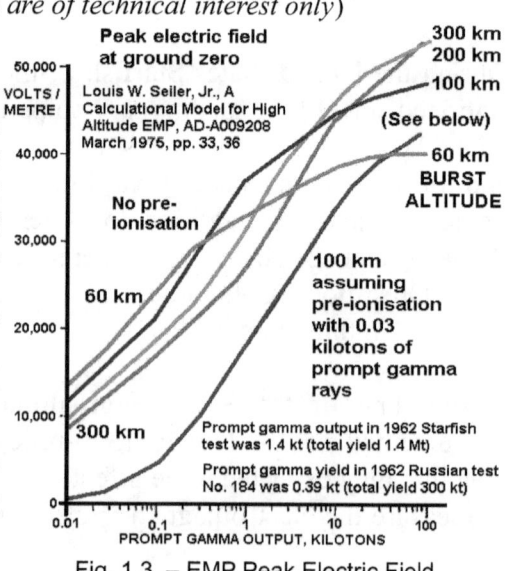

Fig. 1.3 – EMP Peak Electric Field

24

The blue 'pre-ionization' curve in Fig. 1.3 applies where gamma and X-rays from the weapon's primary stage ionizes the atmosphere, making it electrically conductive before the main pulse from the thermonuclear stage. The pre-ionization can literally short out part of the final EMP. Note: Fig. 1.3 also shows that the ground-based electric-field strength approximates 50 kV/m, a value used in MIL-STDs and in the next section, for calculating the 80-dB shielding effectiveness requirement for EMP protection.

1.3-U.S. Government EMP Preparations

The U.S. Department of Defense (DoD), portions of the Pentagon, the Intelligence Agencies and other selected government office buildings and site locations have had their infrastructure EMP hardened. Tens of billions of dollars have been so spent to insure the government can still function in the event of a nuclear event. This also applies to most U.S. Navy warships. *With few exceptions, clearly, no such preparedness exists for any civilian sector.*

Threat Assessments Summary

EMP is an instantaneous, intense energy field that can overload or disrupt at a distance numerous electrical systems and high technology microcircuits, which are especially sensitive to power surges. A large scale EMP effect can be produced by a single nuclear explosion detonated high in the atmosphere. A similar, smaller-scale EMP effect can be created using non-nuclear devices with powerful batteries or reactive chemicals. This method is called *High Power Microwave* (HPM). Several nations, including reported sponsors of terrorism, may currently have a capability to use EMP as a weapon for cyber-warfare or cyber-terrorism to disrupt communications and other parts of the U.S. critical infrastructure. Also, some equipment and weapons used by the U.S. military may be vulnerable to the effects of EMP.

The threat of an EMP attack against the United States is hard to assess, but some observers indicate that it is growing along with worldwide access to newer technologies and the proliferation of nuclear weapons. In the past, the threat of mutually assured destruction provided a lasting deterrent against the exchange of multiple high-yield nuclear warheads. However, now even a single, low-yield nuclear explosion high above the United States, or over a battlefield, can produce a large-scale EMP effect that could result in a widespread loss of electronics, but no direct fatalities, and may not necessarily evoke a large nuclear retaliatory strike by the U.S. military. This, coupled with published articles discussing the vulnerability of U.S. critical infrastructure control systems, and some U.S. military battlefield systems to the effects of EMP, may create a new incentive for other countries to rapidly develop or acquire a nuclear capability.

Policy issues raised by this threat include (1) what is the United States doing to protect civilian critical infrastructure systems against the threat of EMP, (2) could the U.S. military be affected if an EMP attack is directed against the U.S. Civilian infrastructure, (3) are other nations now encouraged by U.S. vulnerabilities to develop or acquire nuclear weapons, and (4) how likely are terrorist organizations to launch a smaller-scale EMP attack against the United States?

Policy Analysis

Private Sector Vulnerability

What is the United States doing to protect critical infrastructure systems against the threat of electromagnetic pulse? What is the appropriate response from the United States to a nuclear EMP attack, where there may be widespread damage to electronics, but relatively little, or possibly no loss of life as a direct result? How could the United States determine which nation launched a EMP attack? After experiencing a EMP effect, the United States may retain its capability to use strategic weapons for nuclear retaliation, but will the U.S. industrial base and critical infrastructure be crippled and incapable of supporting a sustained military campaign? During such time, would the United States be capable of making an effective response should other nations chose to make military advances in other parts of the world?

Some assert that little has been done by the private sector to protect against the threat from electromagnetic pulse, and that commercial electronic systems in the United States could be severely damaged by either EMP or smaller-scale HPM. Officials of several U.S. power stations and public utilities have stated that their electrical systems currently have no protection against electromagnetic pulse. However, electric power and telephone utilities have been known to fail as a result of solar storms which cause effects similar to, but less severe than EMP from a nuclear blast. Commercial electronic surge arresters used for lightning strikes reportedly do not clamp fast enough to protect against the instantaneous effects of the electromagnetic pulse, and some also may not have great enough current carrying capacity.

Military Vulnerability

The effects of large-scale EMP have been studied over several years by the Defense Atomic Support Agency, the Defense Nuclear Agency, and the Defense Special Weapons Agency, and is currently being studied by the Defense Threat Reduction Agency. However, the application of the results of these studies has been uneven across military weapons and communications systems. Some analysts state that U.S. strategic military systems (intercontinental ballistic missiles and long-range bombers) may have strong protection against EMP, while most U.S. weapons systems used for the battlefield do not, and that this uneven protection is undoubtedly known to our potential adversaries.

Some analysts reportedly state that limited testing has shown that modern commercial equipment may be surprisingly resistant to the effects of electromagnetic pulse, and in addition to the SCAMP system, some military systems using commercial equipment are retrofitted to increase resistance to EMP. However, there is disagreement among observers about whether test procedures used by the U.S. military may have been flawed, leading to erroneous conclusions about the effects of electromagnetic pulse on commercial electronics.

The U.S. military has adopted a policy where possibly vulnerable commercial electronic equipment is now used extensively in support of complex U.S. weapons systems. For example, a large percentage of U.S. military communications during Operation Iraqi Freedom was reportedly carried by commercial satellites, and much military administrative information is currently routed through the civilian Internet. Many commercial communications satellites, particularly those in low earth orbit, reportedly may degrade or cease to function shortly after a high altitude nuclear explosion. However, some observers believe that possible EMP and HPM vulnerabilities of military information systems are outweighed by the benefits gained through access to innovative technology and increased communications flexibility that come from using state-of-the-art electronics and from maintaining connections to the civilian Internet and satellite systems.

New Incentive to Develop a Nuclear Capability

A single nuclear device exploded at an appropriate altitude above the continental United States could possibly affect U.S. industrial capacity, economic stability, and military effectiveness. Does knowledge of this vulnerability, combined with the proliferation of nuclear technology, provide a newer incentive for potential adversaries to develop or acquire a nuclear weapons capability? Will countries now view the development and acquisition of nuclear weapons, even a small arsenal, as a strategy for cyber-warfare?

During the Cold War, an EMP attack was viewed as the first step of a nuclear exchange involving many warheads, but the threat of mutually assured destruction provided a lasting deterrent. Today, the proliferation of nuclear technology makes the threat of EMP more difficult to assess. Would the leader of a rogue state be motivated to use a small nuclear arsenal to launch a crippling EMP strike against the United States, with no resulting fatalities, if it believed the U.S. would likely not retaliate with a nuclear salvo that would destroy thousands, or millions of innocent people? Would a EMP strike over a disputed area during a regional conflict be seen as a way to defeat the communications links and network centric warfare capability of the U.S. military, and gain maximum battlefield advantage from an existing supply of smaller nuclear warheads?

Terrorist Use of HPM

A smaller-scale HPM weapon requires a relatively simple design, and can be built using electrical materials and chemical explosives that are easy to obtain. It is estimated that a limited-range suitcase-sized HPM weapon could be constructed for much less than $2,000, and is within the capability of almost any nation, and perhaps many terrorist organizations. Recently, DoD recruited a scientist to create two small HPM weapons for testing using only commercially available electrical components, such as ordinary spark plugs and coils. One device was developed that could be broken down into two parcels so it could be shipped by regular mail, for example, from one terrorist to another. The second HPM device was constructed to fit inside a small vehicle. Aside from specially-trained dogs, experts reportedly say there are no scientific methods that currently allow easy detection of an explosive device hidden in a vehicle or inside a suitcase before it can explode.

It is difficult to assess the threat of a terrorist organization possibly using a smaller-scale HPM weapon against the United States critical infrastructure. It could be argued that an HPM bomb by itself, may not be attractive to terrorists, because its smaller explosion would not be violent enough, and the visible effect would not be as dramatic as a larger, conventional bomb. Also, constructing an HPM device is still somewhat more technically complex than constructing a conventional bomb. However, observers have reported that the leadership of terrorist organizations may increasingly become aware of the growing advantages from an attack launched against U.S. critical information systems. In addition, the use by a terrorist group of a new weapon directed at U.S. information systems would attract widespread media attention, and may motivate other rival groups to follow along a new pathway.

Additionally, the explosives used in a smaller, or suitcase-sized HPM device could simultaneously be used to disperse radioactive materials, making it a so-called "dirty bomb". This combination would offer a possible two-for-one effect, where the dispersed radioactive materials could generate immediate near-panic, while the HPM-damaged computers might not be noticed until days later. This potential double effect could improve the attractiveness of using an HPM device as a terrorist weapon.

Champ, Robo Planes, Black Dart

Drones of the type used in Afghanistan with high explosives to take out terrorist and special targets can be equipped, instead, with HPM to burn-out enemy electronics on the ground. They radiate shorts bursts of microwave energy and their effective range is limited (classified) to, perhaps, one mile.

Since an adversary can use similar "robo planes", they must be disabled by ground-based or air platforms capable of knocking out the robo planes (burns out their electronics) with similar bursts of microwaves. Project Black Dart and others address these issues. However, they are beyond the scope of this book.

Objections

EMP and HPM energy weapons primarily damage electronic systems, with little or no direct effect on humans, however, these effects may also be difficult to limit or control. EMP or HPM energy fields, as they instantly spread outward, may also affect nearby hospital equipment or personal medical devices, such as pacemakers, and may damage critical electronic systems throughout other parts of the surrounding civilian infrastructure. For this reason, some international organizations may object to the development and use of EMP or HPM as weapons.

Legislative Activity

In 1997, the House National Security Committee held a hearing on the Threat Posed by Electromagnetic Pulse (EMP) to U.S. Military Systems and Civil Infrastructure, and in 1999, the House Military Research and Development Subcommittee held a hearing on the potential threats to United States civilian and military systems from an electromagnetic pulse attack.

A Commission to Assess the Threat from High Altitude Electro-magnetic Pulse was established by Congress in FY2001 after several experts expressed concern that the U.S. critical infrastructure and military were vulnerable to EMP attack. Seven of the Commission members were appointed by the Secretary of Defense and two by the Director of the Federal Emergency Management Agency. On July 22, 2004, the Commission presented a report to the House Armed Services Committee, stating that EMP is capable of a causing catastrophe for the nation. However, the report, which focuses mainly on the effects of EMP and not necessarily on HPM, also states that such a catastrophe can be prevented by following recommendations made by the Commission.

Testimony during the presentation raised questions, such as: (1) how would the United States respond to a limited EMP attack against the U.S. homeland or against U.S. forces, where there is loss of technology, but no loss of life; (2) does the current lack of U.S. preparedness invite adversaries to plan and attempt an EMP attack; and (3) are the long-term effects of a successful EMP attack, leading to possible widespread starvation and population reduction, potentially more devastating to the U.S. homeland than an attack by surface nuclear weapons?

References 2:

1. Michael Sirak, "U.S. vulnerable to EMP Attack,"Janes's Defense Weekly, July 26, 2004.

2. Daniel G. DuPont, "Panel Says Society At Great Risk From Electromagnetic Pulse Attack,"Inside the Pentagon, July 15, 2004, p.1.

3. U.S. Congress, House Armed Services Committee, Committee Hearing on Commission to Assess the Threat to the United States from Electromagnetic Pulse Attack, July 22, 2004.

4. Stanley Jakarta, statement before the House Military Research and Development Subcommittee, hearing on EMP Threats to the U.S. Military and Civilian Infrastructure, Oct. 7, 1999.

Lowell Wood, Statement before the House Military Research and Development Subcommittee, hearing on EMP Threats to the U.S.

5.Military and Civilian Infrastructure, Oct. 7, 1999; Jack Spencer, "America's Vulnerability to a Different Nuclear Threat: An Electromagnetic Pulse," The Heritage Foundation Backgrounder, No.1372, May 26, 2000, p.6.; and Carlo Kopp, "The Electromagnetic Bomb: A Weapon of Electrical Mass Destruction," Air and Space Power,1993, [http://www.airpower.maxwell.af.mil/airchronicles/ kopp/ apjemp.html].

6.A nuclear explosion produces gamma rays, which interact with air molecules in a process called the Compton effect. Electrons are scattered at high energies which ionizes the atmosphere, generating a powerful electrical field. This EMP effect is strongest at altitudes above 30,000m, and lasts so briefly that current cannot start flowing through a human body to cause harm to people.
[http://www.physics.northwestern.edu/classes/2001Fall/Phyx1352/19/emp.htm].

7. The Federation of American Scientists, "Nuclear Weapons EMP Effects,"
[http://www.fas.org/nuke/intro/nuke/emp.htm].

8. A Flux Compression Generator consists of explosives packed inside a cylinder, all of which is contained within a cylindrical copper coil structure. The explosive is detonated from rear to front, causing the tube to flare in a wave that touches the copper coil, which produces a moving short circuit. This compresses the magnetic field and creates an electromagnetic pulse that is emitted from the front end, which is then directed by a special focusing antenna.
[http://www.physics.northwestern.edu/classes/2001Fall/Phyx1352/19/emp.htm].

9. Dr. Robert C. Harney, Naval Postgraduate School, Apr. 12, 2004, personal communication.

10. Victorino Matus, "Dropping the E-bomb," The Weekly Standard, Feb. 2, 2003.

11. William Graham, Electromagnetic Pulse Threats to U.S. Military and Civilian Infrastructure, hearing before the Military Research and Development Subcommittee, House Armed Services Committee, Oct. 7, 1999; and Carlo Kopp, "The Electromagnetic Bomb: A Weapon of Electrical Mass Destruction," Air and Space Power, 1993, at
[http://www.airpower.maxwell.af.mil/airchronicles/kopp/apjemp.html].

12. [http://www.physics.northwestern.edu/classes/2001Fall/Phyx135-2/19/emp.htm].

13. Experts may disagree on whether the damaging effects of HPM actually diminish following the familiar inverse-square-of-the-distance rule. Michael Abrams, "The Dawn of the E-Bomb," IEEE Spectrum, Nov. 2003,
[http://www.spectrum.ieee.org/WEBONLY/publicfeature/nov03/1103ebom.html].

Some experts state that the severity of HEMP effect depends largely on the bomb design, so a specially-designed low yield bomb may pose a larger HEMP threat than a high yield bomb. Lowell Wood, statement before the House Research and Development Subcommittee, hearing on EMP Threats to the U.S. Military and Civilian Infrastructure, Oct. 7, 1999.

14. Victorino Matus, "Dropping the E-bomb," The Weekly Standard, Feb. 2, 2003.

15. The Federation of American Scientists, "Nuclear Weapons EMP Effects," [http://www.fas.org/nuke/intro/nuke/emp.htm], and Report of the Commission to Assess the Threat to the United States from Electromagnetic Pulse (EMP) Attack, Vol.1: Executive Report 2004, p.5.

16. Kenneth R. Timmerman, "U.S. Threatened with EMP Attack," Insight on the News, May 28, 2001.

17. Lowell Wood, statement before the House Research and Development Subcommittee, hearing on EMP Threats to the U.S. Military and Civilian Infrastructure, Oct. 7, 1999.

18. Electrical systems connected to any wire or line that can act as an antenna may be disrupted. [http://www.physics.northwestern.edu/classes/2001Fall/Phyx135-2/19/emp.htm]. Army Training Manual 5-692-2, April 15, 2001, "Maintenance of Mechanical and Electrical Equipment at Command, Control, Communications, Computers, Intelligence, Surveillance, and Reconnaissance (C4ISR) Facilities, HEMP Protection Systems," Chapter 27, [http://www.usace.army.mil/publications/armytm/tm5-692-2/chap27VOL-2.pdf].

19. Lowell Wood, statement before the House Research and Development Subcommittee, hearing on EMP Threats to the U.S. Military and Civilian Infrastructure, Oct. 7, 1999.

20. Associated Press, "Experts Cite Electromagnetic Pulse as Terrorist Threat,"Las Vegas Review-Journal, Oct. 3, 2001.

21. Michael Abrams, "The Dawn of the E-Bomb,"IEEE Spectrum Online, Nov. 2003, [http://www.spectrum.ieee.org/WEBONLY/publicfeature/nov03/1103ebom.html].

Will Dunham, "U.S. May Debut Secret Microwave Weapon versus Iraq," Reuters, Feb. 2, 2003, [http://www.globalsecurity.org/org/news/2003/030202-ebomb01.htm].

1.4 U.S. Government EMP Specifications and Standards

Department of Defense MIL-STD-188-125-1 establishes minimum requirements and design objectives for high-altitude electromagnetic pulse (EMP) hardening of fixed ground-based facilities that perform critical, time-urgent command, control, communications, computer, and intelligence (C4I) missions. The standard prescribes minimum performance requirements for low-risk protection from mission-aborting damage or upset due to HEMP environments. It also addresses minimum testing requirements for demonstrating that prescribed performance has been achieved and for verifying that the installed protection subsystem provides the operationally required hardness for the completed facility.

Covered fixed ground-based facility types include subscriber terminals and data processing centers, transmitting and receiving communications stations, and relay facilities. Use of the standard for EMP protection of other ground-based communications-electronics facilities that require EMP hardening is also encouraged. The standard applies to both new construction and retrofit of existing facilities. Only local portions of facility interconnects are addressed. The standard implicitly assumes that survivable long-haul communications paths, fiber optic links, or other hardened interconnects between facilities will be provided as required for mission accomplishment. Uniform application of MIL-STD-188-125-1 requirements ensures balanced EMP hardening for all critical facilities in a network. *(It does not follow that these and other DoD specifications are properly applicable to the commercial world).*

MIL-STD-188-125-1 provides over 100 pages of detailed information on protecting ground-based facilities and equipment from the effects of high altitude EMP. In March 1999, the latest version of the US Department of Defense, MIL-STD-199-125-2 Interface Standard was issued. It is entitled, "High-Altitude, Electromagnetic Pulse (EMP) Protection for Ground Base C41 Facilities Performing Critical Time-urgent Missions, Part 2, Transportable Systems, AMSC N/A Area TCSS." The Forward states:

1.-This military standard is approved for use by all Departments and Agencies of the Department of Defense (DoD).

2.-Originally, Military Standard 188 (MIL-STD-188) covered technical standards for tactical and long-haul communications, but later evolved through revisions (MIL-STD-188A, MIL-STD-188B) into a document applicable to tactical communications only (MIL-STD-188C).

3. The Defense Information Systems Agency (DISA) published DISA circulars (DISAC) promulgating standards and engineering criteria applicable to the long-haul Defense Communication System and to the technical support of the National Military Command System.

4. As a result of a Joint Chiefs of Staff action, standards for all military communications are now being published in a MIL-STD-188 series of documents. The MIL-STD-188 series is subdivided into a MIL-STD-188-100

series, covering common standards for tactical and long-haul communications; a MIL-STD-188-200 series, covering standards for tactical communications only; and a MIL-STD-188-300 series, covering standards for long-haul communications only. Emphasis is being placed on developing common standards for tactical and long-haul communications, published in the MIL-STD-188-100 series.

5. This two-part document contains technical requirements and design objectives for EMP protection of ground-based systems and facilities that are nodes in EMP-hardened networks for performing critical and time-urgent command, control, communications, computer, and intelligence (C4I) missions. Part 1 of the document addresses EMP hardening for fixed facilities; this Part 2 addresses transportable systems. The requirements are stringent, in order to avoid both damage and functional upsets that prevent mission accomplishment within operationally prescribed timelines. The standards apply uniformly to all systems and facilities in the end-to-end chain, since disruption of a single node may result in network failure.

6. Performance, acceptance test, and verification test requirements are contained in the body of the standard. EMP-unique acceptance and verification test techniques are provided in their Appendices A, B, C, and D.

7. Implementation of MIL-STD-188-125-1 is supported by MIL-HDBK-423, "EMP Protection for Fixed and Transportable, Ground-Based Facilities. Among the EMP hardening tools and techniques are surge suppressors, shielding, and filtering and their grounding. These are discussed and applied in Chapter 6,

In summary, current High Altitude Electromagnetic Pulse (EMP)) Standards:

MIL-STD-188-125-1: *High Altitude Electromagnetic Pulse (EMP) Protection for Ground-Based C41 Facilities Performing Critical, Time Urgent Missions, Part 1, Fixed Facilities.*

MIL-STD-188-125-2: *High Altitude Electromagnetic Pulse (EMP) Protection for Ground-Based C41 Facilities Performing Critical, Time Urgent Missions, Part 2, Transportable Systems.*

MIL-HDBK-419: *Grounding, Bonding, and Shielding for Electronic Equipment and Facilities.*

MIL-HDBK-423: *High Altitude Electromagnetic Pulse (EMP) Protection for Fixed and Transportable Ground-Based Facilities, Vol. 1: Fixed Facilities.*

MIL-HDBK-423: High Altitude Electromagnetic Pulse (EMP) Protection for Fixed and Transportable Ground-Based Facilities, Vol. 2: Transportable Facilities.

References 3:

1. Broad, William J. "Nuclear Pulse (I): Awakening to the Chaos Factor," Science. 29 May 1981. Vol. 212, pp. 1009-1012

2. Broad, William J. "Nuclear Pulse (II): Ensuring Delivery of the Doomsday Signal," Science. 5 June 1981. Vol. 212 pp. 1116-1120

3. Broad, William J. "Nuclear Pulse (III): Playing a Wild Card," Science. 12 June 1981. Vol. 212. pp. 1248-1251

4. Bainbridge, K.T., Trinity (Report LA-6300-H), Los Alamos Scientific Laboratory. May 1976. Page 53.

5. Baum, Carl E., From the Electromagnetic Pulse to High-Power Electromagnetics. Proceedings of the IEEE, Vol.80, No. 6, pp. 789-817. June 1992.

6. Baum, Carl E., Reminiscences of High-Power Electromagnetics, IEEE Transactions on Electromagnetic Compatibility. Vol. 49, No. 2. pp. 211-218. May 2007.

7. Vittitoe, Charles N.,Did High-Altitude EMP Cause the Hawaiian Streetlight Incident?. Sandia National Laboratories. June 1989.

8. Longmire, Conrad L., "Fifty Odd Years of EMP", NBC Report, Fall/Winter, 2004. pp. 47-51. U.S. Army Nuclear and Chemical Agency.

9. Dyal, P., Air Force Weapons Laboratory. Report ADA995428. Operation Dominic. Fish Bowl Series. Debris Expansion Experiment. 10 December 1965. Page 15.

10. Rabinowitz, Mario (1987) Effect of the Fast Nuclear Electromagnetic Pulse on the Electric Power Grid Nationwide: A Different View. IEEE Transactions on Power Delivery, PWRD-2. pp. 1199-1222.

11. Longmire, Conrad L. Theoretical Notes - Note 353 - March 1985 - EMP on Honolulu from the Starfish Event. Mission Research Corporation.

12. Stephen Younger, et al. Scientific Collaborations Between Los Alamos and Arzamas-16 Using Explosive-Driven Flux Compression Generators. Los Alamos Science, No. 24, pp. 48-71, 1996.

13. Kompaneets, Aleksandr S., "Radio Emission from an Atomic Explosion". Journal of Experimental and Theoretical Physics, Vol. 35. December, 1958 (Russian-language publication)

14. Gilinsky, Victor. The Kompaneets Model for Radio Emission from a Nuclear Explosion. The Rand Corporation. August, 1964.

15. United States Central Intelligence Agency. National Intelligence Estimate. Number 11-2A-63. "The Soviet Atomic Energy Program." Page 44.

16. Defense Atomic Support Agency. ITR-1660(SAN) (Re-designated ADA369152). Operation Hardtack, Preliminary Report. 23 September 1959. (Also see the Chapter 10 pages on Yucca EMP measurements).

17. Armed Forces Special Weapons Project. Report ITR-1655 (Re-designated ADA322231). Operation of Balloon Carrier for Very-High-Altitude Nuclear Detonation. 25 July 1958.

18. Operation Hardtack High-Altitude Test Film, includes films of the Hardtack-Yucca balloon-launched 1.7 kiloton nuclear test, as well as the Redstone missile launched Hardtack-Teak and Hardtack-Orange 3.8 megaton high-altitude tests.

19. Defense Nuclear Agency. Report DNA 6038F. Operation Hardtack 1958

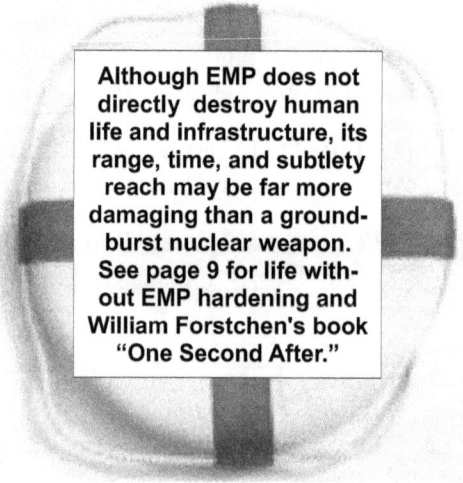

Although EMP does not directly destroy human life and infrastructure, its range, time, and subtlety reach may be far more damaging than a ground-burst nuclear weapon. See page 9 for life without EMP hardening and William Forstchen's book "One Second After."

Chapter 1, Part 2 - Recent EMP Information & What the Future Portends

Chapter 1, Part 2 Overview

The second part of this chapter is a fast forward to the EMP events of recent times plus a look into the future to see what it portends.

EMP Symposia in recent years are identified and summarized. They are a good source of information for the readers. To these add IEEE (Institute of Electrical and Electronic Engineers) EMP publications, primarily generated by the professional group on Electromagnetic Compatibility (EMC) re EMP topics. They contain some EMP material.

The National Geographic magazine has produced a hour-long video on EMP entitled *"Electronic Armageddon: Nuclear Explosion in the Sky."* Also in 1990 a movie was created on EMP which provides some interesting moments of viewing, especially for the non-technical reader (page 40). Others may be found on the Internet and elsewhere.

One section presents an overview of the prevailing national attitude about EMP along with some supporting rationale. Perhaps it is too catastrophic to contemplate, which explains one reason little is done for its prevention in the civil world. This, notwithstanding, some estimates are offered regarding what the EMP future portends. And, there is also a confusion in the minds of many re EMP vs. HEMP. The former is cataclysmic, whereas, the latter provides mostly survivors bereft of electricity and electronics, which can lead to very hostile outcomes.

Finally, a position is offered on what may be regarded as an insurance viewpoint. After all, there exists insurance for almost any situation. You may be surprised by the different forms that insurance for EMP will take, some offering viable solutions to a challenging problem.

Part 2. Recent EMP Info and What the Future Portends

EMP mitigation efforts are gaining attention today from defense contractors and military officials for integration with military communications electronics. During the Cold War, the threat of electromagnetic pulse (EMP) commanded attention and government resources. Today, recognition of the EMP danger is even more important based on our reliance on a sophisticated electronic infrastructure. A survey of 85 arms experts predicted in June 2005,"There is a 70 percent risk of an attack somewhere in the world with a weapon of mass destruction in the next decade". Were this survey made today (2012), the 70 percent estimate would likely be much higher because of publicity about rogue nations and terrorist organizations.

The survey executor, U.S. Senate Foreign Relations Committee Chairman Richard Lugar, describes the weapons of mass destruction (WMD) threat as "real and increasing over time." The Congressional Commission to Assess the Threat to the United States from Electromagnetic Pulse Attack wrote in July 2004:

The high-altitude nuclear weapon generated electromagnetic pulse is one of a small number of threats that has the potential to hold our society seriously at risk and might result in defeat of our military forces.

1.5-Recent EMP Symposia and Meetings

This section, of course, does not disclose or discuss any classified activities in symposia or meetings, as the authors have not held security clearances that would provide access to those meetings in recent years.

EMPact America

EMPact America is a non-profit organization formed by an owner of a food distribution center in upstate New York.

As the owner of a food distribution center, the EMPact America founder realized that after an EMP attack, no food would be distributed. He would have no food products coming into his facility, and nothing would work at his facility in a post-EMP situation anyway. Food distribution would simply grind to a halt.

One of the first significant activities of EMPact America was a major conference on EMP in Niagara Falls, New York in September, 2009. The conference was open to anyone, and most of the speakers were scientific and engineering professionals who were able to explain the EMP threat in a way that anyone could understand. Many speakers also discussed the threat of geomagnetic storms.

The two-day conference was attended by about 700 people from around the world.

The two day event, hosted by *EMPACT America Inc.*, explored the threat, the potential impact on America and options to protect our critical infrastructures and our homes. The conference concluded with a special interactive session to lay out an action plan.

This conference was intended as a watershed event, helping to build a national consensus on the urgency to end our unprecedented vulnerability to a weapon that may already be in the hands of rogue nations or terror groups dedicated to our destruction. As nuclear technology spreads over the 21st century, many additional nations may acquire nuclear EMP capability.

Reports from the Congressional EMP Commission and the National Academy of Sciences characterize electromagnetic pulse as a potentially catastrophic threat.

In a July 21 hearing before a House *Homeland Security* subcommittee, Commission Chairman Dr. William R. Graham testified that EMP ". . . is one of a small number of threats that can hold our society at risk of catastrophic consequences."

Other EMP Symposia

• Other symposia have been conducted in recent years by the US DoD Department of Defense, but are of a classified nature and will not be covered here.

• The Department of Defense (DoD) and the Department of Energy (DOE) announced that the 30th Hardened Electronics and Radiation Technology (HEART) Conference was held in Monterey, CA, March 13 - 16, 2012. The HEART Conference provides a professional forum specifically for classified or sensitive research and development investigations. The HEART Conference and the HEART published proceedings provide a unique, classified and export controlled colloquium for authors who desire the professional recognition associated with fully scientifically-refereed publications. This type of publication is the choice for reporting superior, exclusive results.

Of particular interest are experimental or analytical observations that are new and significant to the design, testing, or economics of hardened systems. Two types of technical papers are invited for consideration in the conference program. The first is classified up to the level of Secret Restricted Data (no CNWDI); the second is unclassified but restricted by export control (ITAR). Papers of special interest, that are not classified or ITAR, will be considered on a case-by-case basis. Attendees must have security clearances certified by DoD and DOE officials to attend the classified sessions. ITAR attendees must be US citizens or U.S. resident aliens with a Permanent Resident Card (Green Card).

Based on the positive reception at previous conferences, demonstration space is provided for those who are developing desktop computer programs in all radiation environment effects, and all related topic areas. The HEART Conference will continue to provide a forum for research and development investigations in space radiation, nuclear, electromagnetic pulse phenomena, and EMP/HPM effects on systems and subsystems. The final program will have a balance of electro-magnetics and solid-state physics phenomena. Papers on new and emerging technologies in related fields are also encouraged.

The technical program will consist of several technical sessions, a half day short-course tutorial, and invited papers. Current research and development results are solicited for presentation.

Solar-PV Conferences

As explained in subsequent chapters, solar-PV plays a significant role in the EMP hardening survival. Several global conferences are planned for 2013 as listed in the table on the next page. Notice that 10 different countries are host. Details may be found on the Internet. Check *"ENF Solar Directory"*.

Solar-related Conferences and Symposia, First half 2013		
Conference or Symposium Name	Where?	When?
PV America East 2013	Philadelphia, PA, USA	5-7 February
Solaire Expo 2013	Marrakech, Morocco	13-16 February
The 2013 PV Euro-Amer, Rollout Conference	Atlanta, GA. USA	25-27 February
International PV Trade Mission	Mexico	25 Feb – 1 Mar
All About Energy 2013	Fortaleza, Brazil	11-13 March
PV System Technology Forum – EU 2013	Dusseldorf, Germany	19-20 March
Solar 2013	Atlanta, GA. USA	13-16 April
POWER-GEN India & Central Asia 2013	Mumbai, India	6-8 May
Greenbuild Expo, 2013	Manchester, UK	8-9 May
PV America West 2013	San Diego, CA, USA	14-16 May
Electric Power 2013	Chicago, IL, USA	14-16 May
Solar 2013 Conference & Expo	Melbourne, Australia	23-24 May
The 24th Annual Energy Fair	Custer, WI, USA	21-23 June

1.6- EMP Publications

Four organizational bodies are a source of EMP publications although EMP is less than 3% of their EMC/EMI output:

• IEEE = Institute of Electrical Engineers, EMC Society Professional Group: At over 400,000 members in 160 countries, the IEEE is the world's largest professional association dedicated to advancing technological innovation and excellence for the benefit of humanity. IEEE and its members inspire a global community through IEEE's highly cited publications, conferences, technology standards, and professional and educational materials.

Because of its wide span of topics covered, IEEE is divided into 38 professional society groups, of which the *Electromagnetic Compatibility Group* (4,000 members) is the one primarily tasked to address HEMP situations. These members are the ones who have provided most of the hardware and services support to the DoD HEMP hardening the U.S. Navy, selected other vehicles and selected office buildings.

The IEEE EMC Society has 71 chapters spread around the world at major population centers to facilitate advancing technological progress. The EMC Society publishes technical papers quarterly. It is the principal source of HEMP support.

• IEC = International Electrotechnical Commission: a non-profit, non-governmental international standards organization that prepares and

41

publishes International Standards for all electrical, electronic and related technologies – collectively known as "electrotechnology". IEC standards cover a vast range of technologies from power generation, transmission and distribution to home appliances and office equipment, semiconductors, fiber optics, batteries, solar energy, nanotechnology and marine energy as well as many others. The IEC also manages three global conformity assessment systems that certify whether equipment, system or components conform to its International Standards.

The IEC charter embraces all electro-technologies including energy production and distribution, electronics, magnetics and electromagnetics, electroacoustics, multimedia and telecommunication, as well as associated general disciplines such as terminology and symbols, electromagnetic compatibility (by its Advisory Committee on Electromagnetic Compatibility. ACEC), measurement and performance, dependability, design and development, safety and the environment.

• CISPR = Comité International Spécial des Perturbations Radio-électriques (International Committee for Radio Frequency Interference) CISPR was founded in 1934 to set standards for controlling electromagnetic interference in electrical and electronic devices, and is a part of the International Electrotechnical Commission (IEC).

HEMP-related publications issued by the above organizations are on the Internet. Some of them are listed in the references of the preceding chapter

1.7 National Geographic and Other EMP Videos

Some readers may enjoy one or more of the following videos:

On Youtube, and available on DVD from the National Geographic Channel web site: Electronic Armageddon: Nuclear Explosion in the Sky.

YouTube has a number of short (approximately 3 minutes) videos on EMP/HEMP ranging from excellent quality to horrible.

A New York Times, best seller novel, "One Second After" by William R. Forestchen, is a fascinating, but scary description re life after a HEMP incident in the United States.

The publisher of this book, Renewable Energy Creation, LLC, is developing various Slide Shows on EMP. Check the web site www.emp-safeguard.com for announcements.

1.8 Prevailing EMP Attitudes and Rationale

It has been 67 years since the end of WWII, when the Hiroshima and Nagasaki atomic bombs were dropped on Japan by a B-29. No atomic or nuclear weapons have been dropped since, except for test purposes described in chapter 1. It has been 23 years since the Brandenburg Gate in Berlin came down and shortly thereafter when the Soviet Union was disbanded into Russia and the other 12 former Soviet republics. No USSR-USA nuclear weapons were used during the Cold War of 42 years. Some are lost in inventory checks and speculations with respect to their whereabouts raise more questions.

Even though USA-USSR had more than 40,000 nuclear warheads between them at the peak in 1985, it seems miraculous that not one was intentionally or accidentally used between the countries. This becomes especially impressive when it is remembered that 107 nuclear warheads were missing when Russia took inventory in Armenia and Azerbaijan following the breakup of the Soviet Union in 1991. Where did these missing weapons go and in whose hands do they exist today? When and where may one or more show up tomorrow, next month or next year? We continue to live on borrowed time. The old Testament in the Bible clearly shows that there have been wars between nations since 2,000 B.C. – for over 4,000 years. Throughout history, it has been man's nature to battle with his contemporaries in many forms of conflicts. In that respect, little has changed. It seems that greed, corruption and power never end.

We have become complacent about the prospects of a nuclear holocaust, after all it has been 68 years since nuclear attack on Nagasaki. The U.S. military realizes that there is a significant chance of an EMP attack, and that they need to be ready for it at any time. That's why the much of the U.S. Navy and Air Force has become EMP-hardened along with scores of selected government buildings. The military services understand the nature of EMP. Meanwhile the remaining 99+% of commercial, industrial, residential and remaining government buildings and other infrastructure are completely vulnerable to any high-altitude nuclear attack. The civilian sector seems to think of EMP as a mystical science fiction concept, while the military knows that it is something that is very real. Nuclear EMP test detonations have actually happened. Nuclear EMP is no more of a figment of imagination than radio signals and electricity are figments of the imagination. EMP could happen as a hostile attack on a continent at any time.

There would seem to make little sense to EMP harden any of the remaining 99+% non-government infrastructure unless a redundant source of electricity can be made available other than batteries and engine-driven generators whose replacement fuel supply is limited and would become of questionable availability following a EMP incident. This suggests the survival need for renewable energy in general and solar-PV in particular. Perhaps, this becomes a more compelling reason for adding solar than the pure economy and environmental aspects of going green considered today. Solar rooftop and other solar-related installations are discussed in nearly all later chapters.

1.9 EMP Protection: Future Guesstimates

The previous section provides the backdrop for EMP protection of all manner of buildings because solar is expected to be on par with fossil fuel replacements (coal, oil and gas) for generating electricity by 2020. Because solar is only available at daytime and with minimum cloud cover, the solar backup of batteries and generators are discussed in chapter 7 in off-the-grid living. One of the benefits in southeast California (for example, the Mojave Desert) of the large windmill farms is that they work through the night when solar is inoperative. For EMP survival, both must have EMP protection.

Solar installations, whether rooftop or solar farms, will have a break even (over the electric-grid provided electricity) in the first few to several years. Later, the ROI (Return on Investment) is shown to be impressive.

Although EMP does not directly destroy human life and infrastructure, its range, time, and subtle reach may be far more damaging than a ground-burst nuclear weapon. See page 9 for life without EMP hardening and William Forstchen's book "One Second After."

Chapter 2

A Contrast of Post EMP Lifestyles - Living vs. Surviving

Chapter 2 Overview

This chapter sets the stage for describing many things required for physical realizability of EMP protection in the rest of the book.

Although the U.S. government has EMP-hardened the military sector, intelligence community and other selected office buildings, essentially nothing has been done in the civil sector of residential, commercial, industrial, and public utilities. This seems ironic since employees of DoD (Department of Defense), DHS, DoE, DoT and others, nearly all live off premises, shop in malls, shop pharmacies, buy gasoline and other products from unprotected facilities. So, in essence, the U.S. remains unprotected from EMP.

The Senate Committee on Homeland Security and other pro-EMP protection organizations are perceived to be frustrated from lack of timely U.S. EMP protection action. Some blame it on American apathy, preoccupation with recession, and unemployment. EMP survivalists, have taken surviving into their own hands by default, some even decades ago. Therefore, this chapter addresses a few examples of strategies, tactics, planning and implementation options.

Post-EMP lifestyle is compared and contrasted with Pre-EMP lifestyle. While EMP survivalists plan to move to the country, storing water, 25-year shelf-life, freeze-dried food, medications, guns ammo, and barter items, they fail the lifestyle test. On the first day of the EMP event, they lose their job, die from failure of the dialysis machine, lack of gasoline and scores of certain provisions. But, their focus is surviving, defined as a bridge from a disaster to the day things are "restored". However, EMP "things" may never restore.

Other interim lifestyle options are offered at HOA, hamlet, village and town levels for EMP protection, but begin to fail at the city levels, where survival crimes increase manifold. Due to highrise structure limitations, the starving are stalking the survivalists for their own surviving, since goods replenishment lags need in bigger cities.

This chapter contains many ideas on how to improve lifestyle with EMP protected town size, a landing strip and railroad siding for replenishment augmentation. 64 million USA inhabitants are members of Home Owners Association., who could make protection happen at the residential level.

A peak at the cost to orchestrate Phase-1 of a two-phase scenario is presented for each of four EMP-protected classes. The total cost is $1.7 billion and 4,200 jobs are generated. Near the 8-year implementation end of Phase-2 EMP protection, an outcome of Phase-1, 7,000,000 jobs will have been created, protecting most U.S. homes and buildings and selected infrastructure.

2.1- The Pros and Cons of the EMP Survivalists

In searching the Internet under the keywords, "EMP Protection", more than half of the listings are from EMP Survivalists. They are selling freeze-dried, 25-year shelf life food, Faraday cages (shielded shoe boxes or other metal-foil lined containers) to protect their electronics, backpacks for different applications, books, and much more. Their mission is basic survival.

After a disaster, survival is often thought of as a bridge between lifestyle before the act of nature or man-made event strikes and when restoration of previous lifestyle is achieved. For hurricane Sandy in 2012, this may have taken from days to years depending on where you live. The default elapse time for an EMP Survivalist is about one year – relatively short by some standards.

Just how does the Survivalist score rate when compared to other EMP-protection options. Figure 2.1 suggest three elements of many comparisons.

Fig. 2.1 – Standard of Living after an EMP, assuming the entire local gathering is also EMP protected.

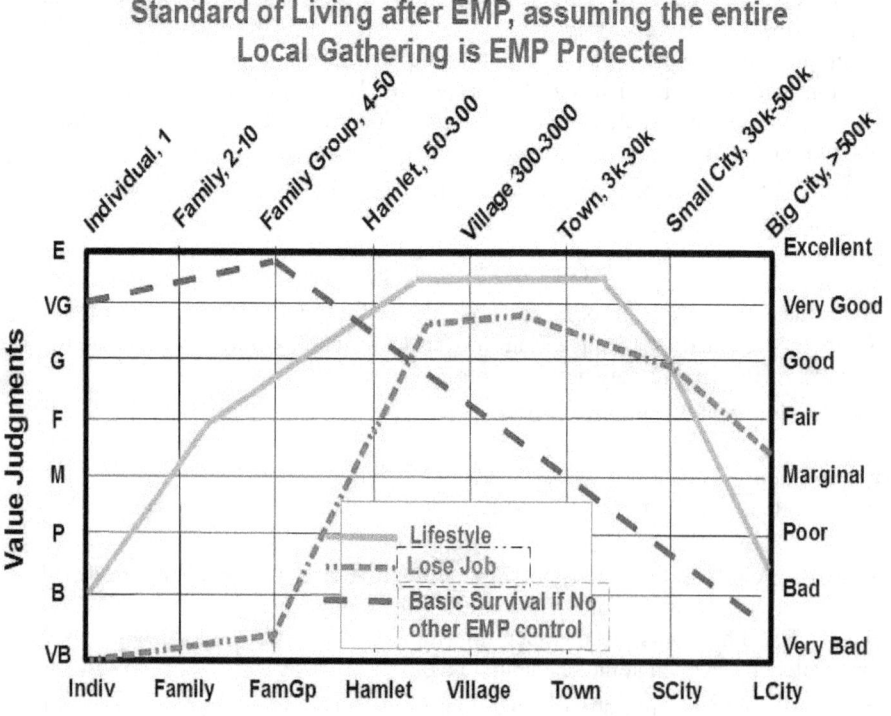

The Surviving group size from an individual to a large city is displayed as the horizontal axis. A value judgment is shown for the vertical axis. It ranges from excellent to very bad. Of course, some value judgments are in the eyes of the beholder (like lifestyle, per se), while others are absolute, like losing your job.

Fig. 2.1 shows that basic surviving is scored very good compared to no other EMP protection. It scores particularly bad in the big city since there is a plethora of starving people ready to break into the homes of anyone storing food. That's why survivalists prefer to move to the country when an EMP event happens.

Fig. 2.1 also shows that one major disadvantage of the EMP survivalists is that he loses his job after an event since nothing else is working anyway. However, if a village or town were EMP protected, and almost all employment is geared to the local events of a city, job retention may approximate 90% after an event. In fact it could theoretically exceed 100% if the town is a manufacturer or it develops products or services in demand from unprotected neighboring areas struck by the same EMP event.

2.2- The Concept of a Four-Tier EMP Protection Plan

Lifestyle (L) following an EMP incident and corresponding cost (C) for EMP protection are the two principal elements to consider, plan for and score EMP protection performance (PP). Therefore, PP = L/C quotient. This PP term will be used in several places including later chapters along with the many facets of EMP considerations.

Lifestyle is rarely addressed in EMP protection literature. Does it not matter? Yet, EMP survivalist will lead a very different lifestyle vs. those in a whole   village or town that is completely EMP protected. For example, following an EMP event, the EMP survivalists individual (or family) loses his job, loses access to shopping stores, hospitals, undertaker, etc. that have all become dysfunctional. Contrast this to an EMP-protected municipality in which almost nothing is lost except uncertainty of when the EMP-protected replenishment vehicles, airstrip or railroad siding may be revisited with more replenishments. Even here, a warehouse can store survivalists freeze-dried and selected canned food, principal   medications, etc. These matters lead to the reason why different tiers of EMP protection are addressed as one strategy of many.

The chapter overview indicates that there are four tiers of EMP protection to be initiated in order to get things started. Except for EMP survivalists, who have started their planning and implementing years ago, the top three tiers may initially be regarded as Phase-1, pilot programs, from which a substantial pragmatic learning experience develops. They are started at the same time and have been selected by location and economy as discussed later. Each tier advisory group has a representative of the other two tiers plus an EMP survivalist, a county Economic Development office participant, and a Chamber of Commerce person since all learn from the progress, failures, and wisdom of each other. Also, as discussed later, financing comes from issuing county and corporate bonds.

To get started, the 3,141 counties in the U.S. are divided into four quartiles of median household income: (1) Tier-1, highest quartile (25% of total US households earn more than $69,000), (2) Tier-2, median quartiles (2nd and 3rd quartiles earn between $22,000 and $69,000 with $39,000 being their geometric midpoint), and (3) Tier-3, 25% of the total US county households earn less than $22,000 per year.

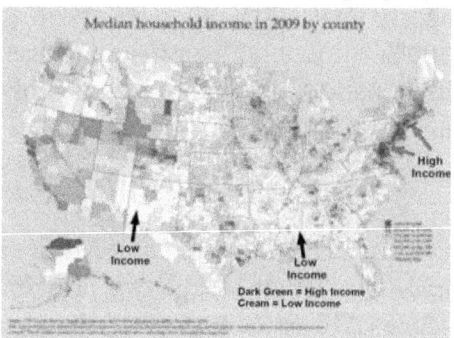

Fig.2.2– County map of USA showing household income divided into five groups

The above data are obtained from the Census Bureau. The U.S. map, Fig. 2.2, shows the household income by darkness of color for each county. The dark colorations are Tier-1 counties and light or no coloration are Tier-3 counties.

The master plan also provides for north and south locations and three geographical regions in the US to be similar in the four-tier EMP protections. This allows for coastal USA exposures, coastal-inland and central regions in order to gain information different from each other's location.

The diagram, Fig. 2.3, illustrates the four tiers of EMP protection. Their assigned names are located at the outer periphery of the four-sides of the square. Just inside the square, the corresponding household income range is listed. As mentioned, Tier-1 counties can afford greater (full) EMP protection and Tier-3 less. Of course, there is allowance for exceptions, not discussed here. One example is that many counties will have poorer sections among the wealthier section locations within a single county.

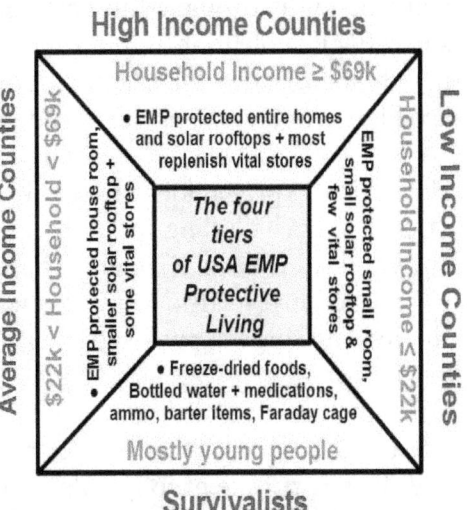

Fig. 2.3: 4-Tier EMP Protection Concept

Fig. 2.3 also has a three-line brief, closest to the center square, to suggest what is covered in their respective tiers. A few details of these remarks are described in the next section.

2.3- How the EMP Protection is Configured

The following discussion addresses some information about each of the four tiers of Fig. 2.3.

Tier-1, High Income Homes, Commercial and Other Buildings

There are 91 million homes in USA and 39 million apartments to shelter it's 314 million population (year 2012). For Phase-1, two-year pilot program, a town of about 10,000 population is selected for the first tier. With an average household size of 2.6 people, this corresponds to 10,000/2.6 = 3,850 homesteads. Of these, 2,700 are detached homes and 1,150 are apartments. Therefore, at the end of Phase-1, about 3 locations x 3,850 = 11.6 thousand Tier-1 homes will be EMP protected. Parenthetically, assuming Phase-2 is completed eight years later, up to 23 million Tier-1 homes in all of the U.S. will be EMP protected with new jobs running into the millions.

Since average new home and new commercial building construction is assumed to be about 2.5% per year, Tier-1 site, 0.025 x 3,850 = 96 new Tier-1 homes per year will be added per site. This is first mentioned now to inform the reader that EMP protection for new home construction is estimated to cost about 65% of that for a retrofit home, since EMP protection is more easily and economically achieved on a new home during construction. Also, the EMP bottom floor shielding problem is overcome as discussed in chapter 5.

Home EMP protection in Tier-1 involves a 100% shielding of the outside skin (including the floor), the details for achievement of which are provided in later chapters. The shielding is bonded and grounded to earth and any lead-in power lines and others (telephone, data) are filtered and surge suppressed as also presented in later chapters.

Since the low end of home size in the high income counties is roughly 4,000 sq. ft. under air, the power required for a solar rooftop is about 10 kW. This is sufficient to handle air conditioning, and hot water loads in addition to the electrical appliances, lighting, computer and peripherals, radio and TV, etc. Of course, the solar rooftop is also shielded and processed as explained in later chapters.

Along with the solar rooftop is both a battery bank of about 30, 12-volt vehicle-type, lead-acid batteries. They provide an energy capacity of about 30 kwh (kilowatt hours), nominally sufficient to handle all night time use and a few overcast days when solar electricity is nearly unproductive. As explained later, the number of batteries is adjusted for greater latitudes and climates having more overcast days.

The cost for a high income county, EMP protected home with protected solar rooftop will range from about $50,000 to over $100,000 for large homes over 10,000 sq. ft. Ignoring inflation, this will reduce by about 30-40% in 10 years by Moore's Law for electronics and quantity production cost reduction. (Moore's Law is the engineering dictim that, among other things, describes the cost reduction of semiconductor products over time.)

The Fig. 2.3 wording speaks of "Larger retail replenishing stores." This term is illustrated in the table of Fig. 2.4. Focus on Column B for the present. Scan down the list of facilities that will have been EMP hardened. It is noted that nearly everything has been protected so that the affluent town as a whole is nearly unaware of an EMP incident. The reason that the word "nearly" is used is that communication and transportation of delivery vehicles, delivering replenishment food, medications and other is essential to survival. This requires that some modes of communication such as satellite and fiber optics are functional. Also, all Tier-1 communities have at least a 4,000 foot metal-mat runway to help ensure vitals replenishment.

Fig. 2.4 – Identification of Principal Resources for survival and living needs.

A Buildings and Topics	B Hi-Income Counties	C Median Counties	D Lo- Income Counties
Shopping Malls	X	X	
Wal-Mart	X	X	
Vehicles	X	X	
Food Stores	X	X	X
EMS Services	X	X	
Pharmacies	X	X	X
Adjacent Farms	X	X	X
Water Utilities	X		
Gas Stations	X	X	
Airport-Trains	X	X	
Manufacturing	X		
Hospitals	X		
Clinics	X	X	X
Weapons-Ammo	X		
Boats-Ships*	X	X	
Marinas*	X		
Commerc. Office	X	X	
Home Depot,-Lowes	X		
Ace Hardware	X	X	X
Hotels/Motels	X		
Schools-College	X	X	
Funeral Services	X	X	
Churches	X	X	
Restaurants	X		
Theaters	X		
Arenas			
Animal Hospital	X	X	

Tier 2 - Median Household Homes, Commercial & Other Buildings

This second tier, median income counties, Column C, Fig. 2.4, has at least one room in a home selected for shielding plus a limited EMP protected solar rooftop of about 3 kW. Unlike the entire home in Tier-1, it is shielded on the *inside* of the room to protect it from all the exposed wiring in the walls (that act as EMP radiation pick up antennas), ceiling lights and duplex outlets that would otherwise severely compromise shielding effectiveness performance. The wiring from the protected solar rooftop goes directly to a small electrical panel *inside* the set-aside room – not to the main electrical control panel used to service the house distribution from the electric grid. Chapter 6 provides more information and details.

The cost for a median income county, EMP-protected room in a home with 3 kW protected solar rooftop (plus supporting solar battery bank and generator) will range from about $15,000 to $35,000 for the larger end of homes approximating 3,500 square feet.

Fig. 2.4 shows that some of the commercial buildings and services are not protected since the county budget is less and other than the main survival retail outlets (for example: Walmart, Home Depot, Walgreen Pharmacy, etc.) the smaller provisioning stores are not included. Naturally, this has a set of very definitive rules as where the dividing line exists, and constitutes a potential political, emotional, and economic problems addressed elsewhere.

Tier 3 - Low Income Homes, Commercial and Other Buildings

EMP Survivalist Provisioning	
Water	**Electric & Electronics**
Bottled	Light appliances
Canned	Flashlights
Well, hand pump	Batteries
Charcoal & sand filtered	Battery charger
Soft drinks	Shortwave radio
Brook or creek	Police Scanners
Food	CB radio
Canned food	Cell phone
Freeze dried	Small Solar
Vegetable garden	Inverter
Cows and pigs	iPad and/or iPod
Deer and squirrel	Microwave oven
Medications	**Computer & Peripherals**
Prescription drugs	**Candles and housing**
Vitamins	**Cooking**
Ointments	Propane heater
Pain killers	Propane storage tank
First aid kit	Charcoal
Splints and bandages	Box oven
Oxygen or other	Tin Can & Dutch ovens
Weapons and Ammo	Eating flatware
Pistols & revolvers	Kitchen utensils
Hunting rifles, shotguns	
Ammunition	
Machete	**Security**
Carving knives	Trip alarms
Bow and arrows	Proximity alarms

Fig. 2.5 – Partial listing of candidate items EMP survivalist will seek to acquire.

This is a continuation of the previous two EMP protected homes in Tier-1 and Tier-2 and provisioning retail stores and other services. Some suggested details are given in Column D of Fig. 2.4.

The selected room for shielding is smaller than for that for Tier-2. Perhaps the smallest bedroom or a pantry or a walk-in clothes closet will work out where the room is shielded and made into basically a large Faraday cage (Faraday and shielded are the same). The solar rooftop consists of three to five, 3'x5' solar panels (about 600 watts to 1,000 watts) used for small lighting, a fan and a small refrigerator. No generator is involved but about five lead-acid vehicle batteries provide the electric backup for night and overcast skies.

The cost for the above limited EMP installation ranges from about $3,500 to $7,500, but becomes less in time as quantity production and installations increase into mid and late 2010s.

Tier Four, EMP Survivalists

This group differs from the above three tiers, in that the EMP survivalist are independently operating. Their motivation to spend time and their own money for EMP protection is in recognition of earlier-mentioned apathy and the fear of the consequences. This has been exacerbated by the fact that the civil sector has done almost nothing and they perceive that there exists little evidence that the civil sector will be doing much anytime soon. Their EMP survival preparation experience has developed a wealth of basic existence know-how.

A browse through the Internet under an "EMP Protection" keywords or related search provides over 100 advertisers and listings of items for sale including a few books. The table in Fig. 2.5 lists most of what may be found with such a search. One of the more comprehensive books the reader may want to buy under $10 is *EMP Survival* by Larry and Cheryl Poole. It may be found on Amazon Books.

2.4- Evaluating Progress: Feedback, Upgrade & Learning

In summary, from the above there are three regions in the US Phase-1, pilot EMP protection design and implementation plan to be achieved in about two years. The first six months is spent on paperwork in formulating specific architectural designs for each of the three economic county levels plus the EMP survivalists. This is nine pilot sub programs (plus another three for Tier-1 for a smaller population discussed later) plus EMP survival lists for retrofitting existing homes. In each of the three tier locations, a few new homes and buildings are added to account for the 2.5% new construction per year. Also, it is remembered that EMP hardening is less expensive for new homes, only achievable during new home construction. These data are also needed for Phase-2.

Fig. 2.6– Map of U.S. showing one of many possible locations for each Phase-1 site

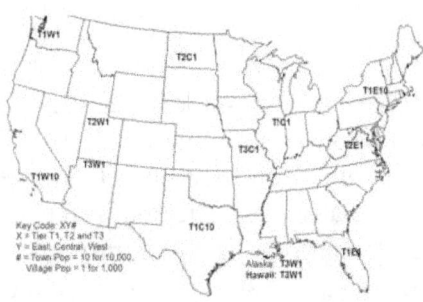

Figure 2.6 is one assigned layout for the pilot program. Obviously, there are political, emotional, economic and other considerations in such an assignment. It must also consider climate, economic and other factors. Then, within these regions for two or more candidates, it may be necessary to select the final choice based on a lottery system.

Progress throughout the two years of Phase-1 will be known to each selected county on a periodic or continuous basis so that all will benefit from the good vs. not-so-good choices with supporting rational. In a different vein, throughout the entire program a number of seminars and webinars will be run nationally and locally so that the non-technical readers as well as professionals will be informed on a national basis. To add to the national interest and participation, there will be cash awards for submission of additional design concepts and implementation aspects by the public at large. After all, there is a wealth of creativity spread throughout the country, and USA must participate and capitalize upon these resources.

Details of the above are presented in a new book being written by Donald R.J. White and affiliates.

2.5- A Peek at Phase 1 Costs , Jobs and Expectations

Discussed in depth in later chapters, this section gives the readers a peek at expectations regarding costs and jobs, first for Phase-1, EMP-Solar, Pilot Project (experiment or study) described in this chapter 1. Then, when carried through Phase 2 - the entire USA by 2023.

Fig. 2.7 – Summary of general EMP tier classification and assignment

A	B	C	D	E	F	G	H
Tier & Location	Population	US Population	Homes #	Cost $	Homes $M	Commercl $M	Total $M
Tier 1, East	10,000	0.0032%	2,700	55,000	149	238	386
Tier 1, Central	10,000	0.0032%	2,700	55,000	149	238	386
Tier 1, West	10,000	0.0032%	2,700	55,000	149	238	386
Tier 1, East	1,000	0.0003%	270	55,000	15	24	39
Tier 1, Central	1,000	0.0003%	270	55,000	15	24	39
Tier 1, West	1,000	0.0003%	270	55,000	15	24	39
Tier 2, East	1,000	0.0003%	270	12,000	3	5	8
Tier 2, Central	1,000	0.0003%	270	12,000	3	5	8
Tier 2, West	1,000	0.0003%	270	12,000	3	5	8
Tier 3, East	1,000	0.0003%	270	4,000	1	2	3
Tier 3, Central	1,000	0.0003%	270	4,000	1	2	3
Tier 3, West	1,000	0.0003%	270	4,000	1	2	3
Totals:	39,000	0.0124%	10,530		503	805	1,308

Figure 2.7, in a spreadsheet form, gives a glimpse of the various element data and their totals on the bottom line. Column B is the customer town or village population involved and Column C is the corresponding population in percent relative to the entire USA population of 314 million as of 2012. Column D is the approximate number of homes involved and E is their rough cost per average home. Column F is the total home cost in units of million dollars (M) for the Tier and location shown in Column A.

The electric utility industry reports that 38% of their electricity load (users) are residential and 62% is a combination of commercial plus industrial. The latter is 62/38 = 1.6 times the residential load. So, Column G is 1.6 times the amount of column F. Finally, Column H is a total of both Columns F and G. Remember, the numbers here are rough since there are many expenses and variables involved and the lower cost of government participation and support has not yet been added.

Figure. 2.7 shows that the cost for Phase-1 over the first two years is about $1.3 billion (total at bottom in column H). This is spread over 12 counties with the first three, each serving a town population of 10,000 which is the highest cost at about $386 million. The average population of all 3,141 US counties is ($314 million USA/ 3,141 counties) = about 100,000 people/county. Their (U.S. census) estimated net worth is about $55 trillion/3,141 counties = average of $17 billion/county. Highest income counties may approximate $40 billion and the lower income about $4B. For a high-income county of 100,000 population, $40B net worth corresponds to $400,000 per person.

The $386 M of Column H, town in Fig. 2.7 is spread over a 100,000 county population and amounts to about $3,800 per person. However, this is not relevant since, as mentioned earlier, the money comes from the issuance of county and corporate bonds from investors, retirement funds, and annuities. From the previous paragraph, $3,800 is less than 1% of the county per capita wealth; so this is no financial challenge (in other words, it is readily affordable).

Regarding jobs, the $1.3 billion pilot, Phase-1 cost (bottom line, Fig. 2.7, Column H) is the total direct cost exclusive of government participation. Thirty per cent (30)% is arbitrarily added to account for government participation cost, some volunteer time contributed, publicity and education costs, plus items not identified in Fig.2.7. So the $1.3 billion becomes about $1.7 billion cost for Phase-1.

Since $225,000 of money spent back into an economy represents one job created with a $48,000), the number of jobs created from Phase-1 is $1.7B/$2250k = 7,600 job-years (see Fig. 2.8). Because this is spread over two years, this represents 3,800 jobs lasting for two years. For the entire USA, this will approximate the above (7,600/0.000124) or (Jobs/Total, Col. C, Fig.2.6) = 61 million job-years, or averaged over 8 years = 7.6 million U.S. jobs.

The above financial information is admittedly rough, but adequate to get a first order evaluation of the do-ability of Phase-1.

Fig.2.8 - Cumulative Jobs Generated in Millions Since 2013

The rapid rise in employment after end of Phase-1, is the start of major contracts across USA for EMP protecting homes and commercial and industrial buildings

Phase-2

Phase-1

Event Year

--- *Addendum* ---

Economic and Financial Market Loss from EMP Burst

Not mentioned and not addressed earlier is the resulting potential catastrophic economic and financial loss from an EMP burst. On April 23, 2013, the stock market "lost" about $200 billion dollars for a few minutes on a Twitter-hacked rumor of a White House explosion, before correcting. This resulted from many software account trigger, sell-order programs having a loss mitigation, safety "protection action."

Suppose the rumor had been a sudden EMP burst instead, in which the electric grid went down along with all RAM and backup memory that became dysfunctional. Even flash memory was burnt out by being connected to a mother board. All the Stock brokerage houses, banks, insurance companies, retirement funds, annuity records and the like back-up memory was lost. Are your equity records in the hands of others EMP non-protected institutions? Can you prove your loss to the institutions who have also become dysfunctional ? Your life savings!

Chapter 3

EMP Protecting Homes and Small Buildings

Chapter 3 Overview

This chapter starts with a small building of the order of 15,000 square feet (1,393 sq. meters). At the beginning there is no EMP protection and the identification of 40 representative vulnerable electrical and electronic items are identified.

An EMP explosion will couple into each and every victim separately and, in some cases, by multiple coupling paths. The victims are grouped into building housekeeping, electrical appliances, computer related and personal devices. While electrical devices tend to be less vulnerable to EMP than very low power electronics, many newer versions of electrical apparatus have a microprocessor or two built in today.

In later chapters it becomes immediately apparent that the quickest way to protect all items is by shielding the entire building to the extent of 80 dB below 64 MHz (or other EMP profile to be determined). This shielding effectiveness requirement is also developed in this chapter 3.

Remember that an EMP incident will not kill humans, as contrasted to a low-altitude nuclear holocaust, but results in an area in which nearly all the infrastructure is still in place, but without electricity and electronics, and is therefore inoperative. What message and ideas does this convey and what actions does this suggest?

3.1-Typical Non-EMP Hardened Small Building

The building used as an illustrative example is a small, two-floor structure of 15,000 square feet (1,393 sq. meters). It is illustrated in Fig. 3.1 and will be used again in later examples with many additive parts as its complexity unfolds.

The building is assumed to be completely transparent to electromagnetic radiation which means that no shielding is yet

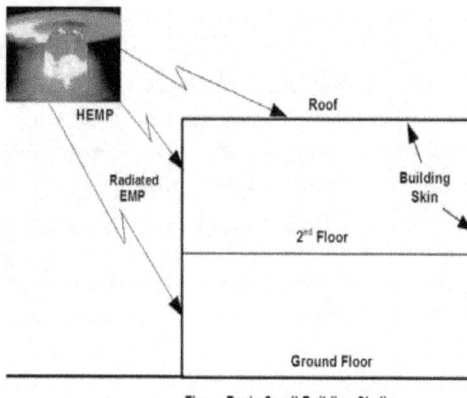

Fig. - Basic Small Building Shell

Fig. 3.1 - Small building surface shell

installed. So, even if the building were disconnected from the electric grid, the many devices within the building are completely exposed and result in a EMP incident burnout.

The building is or should be protected from a lightning strike. This implies that incoming wiring (AC power, telephone lines and possibly control and signal leads) are already surge protected. **Beware**, lightning surge suppressors are far too slow to react to a fast risetime EMP; they are almost useless and may give a false sense of EMP security to the uninformed observer. So, they will need to be updated to become EMP protected.

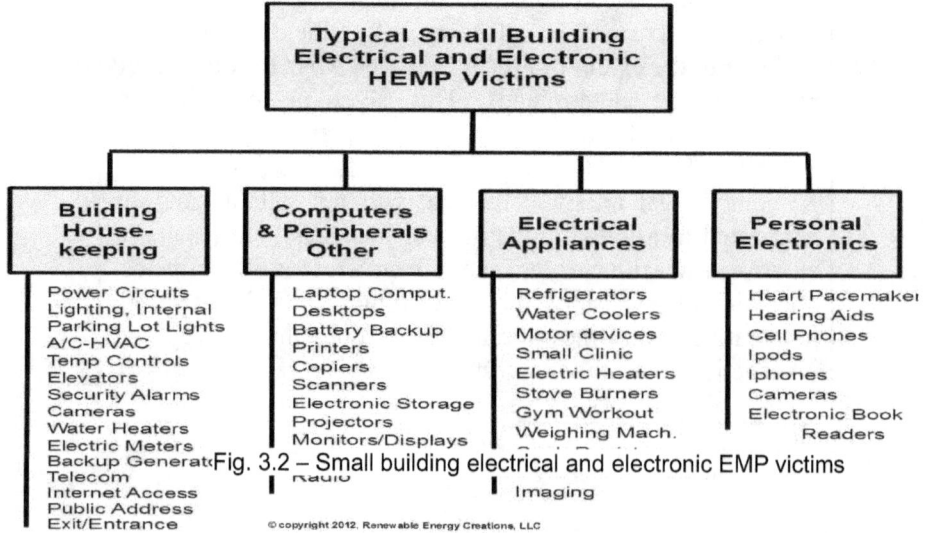

Fig. 3.2 – Small building electrical and electronic EMP victims

3.2 Typical Building EMP Victims

While not shown in the illustration, the building typically contains 80-90% of the victim items listed in Fig. 3.2, on previous page, under the four categories shown. Note, this is the raw building before EMP hardening and, possibly, a solar rooftop installation is added later on, with its battery and generator backup. Each of the 40 listed devices in Fig. 3.2 is subject to being upset, jammed or, more probably, burned out if exposed to an EMP episode. Only if the building is sufficiently shielded and protected will all devices survive. Some options would appear to exist for protecting many of the victims by placing them in a shielded enclosure (room) and/or shielded cabinets or consoles (sometimes called Faraday cages) inside the building. But, even if this is done, 20 house-keeping victims are still exposed without protection outside the shielded room or shielded cabinet(s). The economy lies in shielding the entire building vs. several or many separate shielded rooms within the building.

3.3- Modes of EMP Equipment Coupling

Other than some personal devices in Fig. 3.2, all others have at least two modes of EMP coupling: (1) direct radiation coupling to the victims and (2) coupling through their power lines connected to the AC duplex outlets located in every electrical circuit. Some devices may also have a telephone, Ethernet or other communication cable. Others may be part of a ground loop to constitute a fourth coupling path.

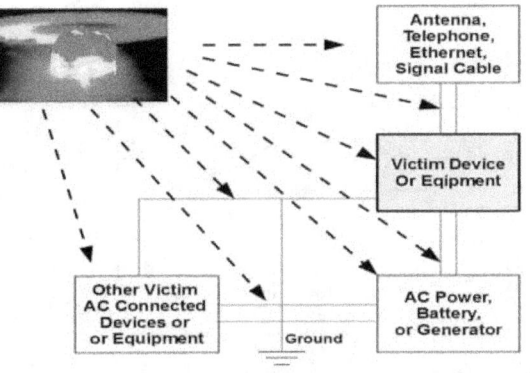

Fig. Eight Radiation Coupling Targets, Some into Cables and Ground Loops

Fig. 3.3 – Eight EMP radiation coupling targets, some into cables and ground loops

In fact many/most AC-powered equipments have 8-10 coupling paths as shown in the victim (depicted in gray in Fig. 3.3. The illustration also shows ground-loop coupling. So, it should be immediately apparent that a shielded building has the highest yield of EMP protection, because once coupling is done into a single device with AC and signal leads, the problem cannot be reversed. Also, the high coupled voltage spikes alone in the cables (some approaching 100,000 volts; see Fig. 3.5 later) will cause burn out of victim power-line filters, surge suppressors and cable insulation arcing.

3.4 EMP Protection Requirements (for technical readers)

As mentioned in an Section 1.4 under MIL-STD-188-125, the EMP pulse in the time domain (X-axis is time) has an amplitude of 50 kV/m, a rise time of 5 nsec and a pulse width at the half amplitude of 150 nsec. This is shown in the insert at the left in Fig. 3.4 and represents the test specifications for EMP testing and compliance.

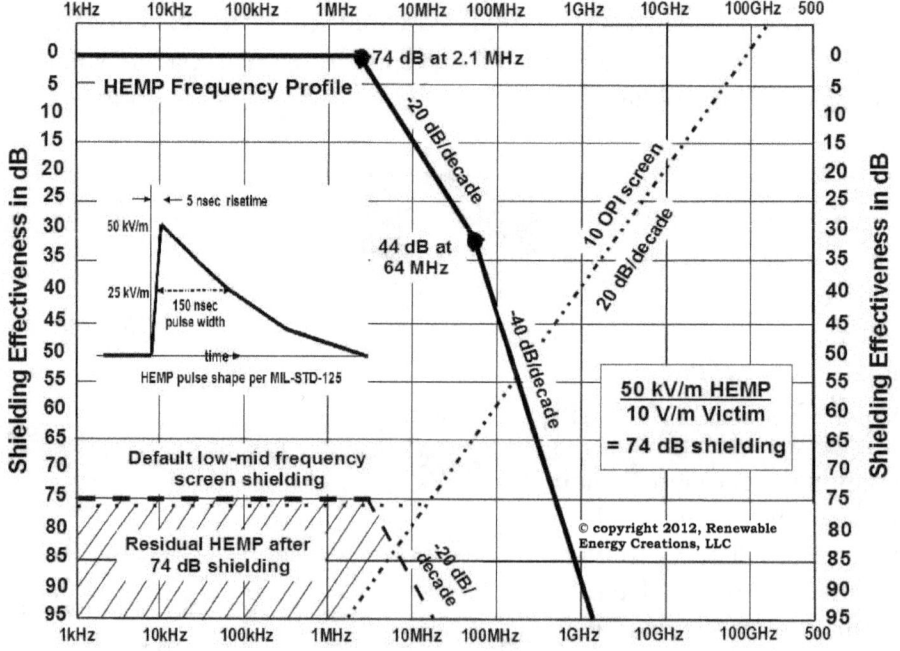

Fig. 3.4 – Shielding effectiveness requirements for EMP protection

Figure 3.4 also depicts the frequency-domain (X-axis is frequency) manifestation of the time-domain pulse just mentioned. It is presumed here that all electrical and electronic devices comply with European Union (CE mark) or other test compliance of radiation susceptibility limits of 10 V/m. (For example, the European Union specifies in EN61000-4-3 that equipment subject to Level 3 requirements shall not be susceptible to a field strength 10 V/m over the frequency range 80 MHz to 1 GHz.

Assuming this to be the case, it follows that the required shielding effectiveness (SE) below 64 MHz (the Second Fourier corner frequency in Fig 3.4) is:

$$SE_{dB} = 20Log_{10}(50,000 \text{ V/m}/10 \text{ V/m}) = 74 \text{ dB} \qquad (3.1)$$

The heavy black line marked "EMP frequency profile" in the upper left is the frequency envelope corresponding to the EMP time-domain pulse shown in the insert. Since a shield has to be identified to compute SE, a 20 OPI (Openings per Inch) screen as selected in Chapter 5 will be used here to cover all windows and solar panels. It degrades with frequency at the rate of 20 dB/decade and stops at 74 dB down due to limitations in the screen wire elements secured to building shielding (a bonding problem) at its periphery.

When the convolution of the EMP pulse-frequency profile is overlaid on the screen attenuation profile, the resultant shielding effectiveness develops. Fig. 3.4 shows the resultant EMP that leaks through in the hatched area in the lower left of the figure. This is the required 74 dB SE. For safety margin and error protection, 6 dB is added to the 74 dB to get a final EMP requirement of 80 dB shielding effectiveness.

Note: Again, this is based on a 50 kV/m EMP threat having a 5 nsec risetime and 150 nsec pulse duration at the 50% pulse height, and victims having a radiated susceptibility of 10 V/m. Referring back to Fig. 1.3, the resulting 80 dB shielding effectiveness would seem to be adequate for all situations meeting this note.

This section has provided the basis for determining the shielding effectiveness, frequency profile and its supporting rationale for buildings discussed in later chapters. If it develops that there should be a change in the culprit EMP radiated-frequency profile or the radiated susceptibility limits of the victims, then Fig. 3.4 must be reworked to get a new updated specification. *Please do not forget this caution.*

3.5 EMP Coupled Voltage into Many Building Victims

To provide a different measure of the EMP ability to burnout or destroy everyday electrical/electronic items used in the workplace and at home, consider the table in Fig. 3.5. The maximum voltage developed in any unprotected device is shown in the last column for the indicated victim in the first column, all corresponding to the waveform shown earlier for Fig. 3.4. That waveform has a 5 nanosecond risetime, a pulse width of 150 nanoseconds, and an amplitude of 50,000 volts/meter.

For example, a 6-ft lamp electric cord develops a maximum end-to-end voltage of 91,000 volts. This can cause insulation break-down or arcing if a grounded metal device is within an inch of another grounded device. A laptop computer with a 15" screen will develop a peak voltage of about 19,000 volts - enough to more than fry the display. A heart pacemaker will develop a peak induced voltage of about 4,600 volts more than 1000 times its internal operating voltage of the microcircuits. What does all this tell us?

Device maximum voltage induced from 50 kV/m HEMP Incident

Victim Identification	Device Maximum Dimension				Induced Voltages
	feet	inches	meters	cm	
Building max. dimension	500	6,000	152	15,240	115,000
House leadin wiring	100	1,200	30	3,048	115,000
Tractor trailer	75	900	23	2,286	115,000
House maximum	50	600	15	1,524	115,000
Pleasure Boat	32	384	10	975	115,000
House circuit wiring	30	360	9	914	115,000
Large auto	18	216	5	549	115,000
6-ft typical lampcord	6	72	2	183	91,440
Desktop Coputer w/ cord	6	72	2	183	91,440
Mwave oven with cord	4	48	1	122	60,960
Laptop Computer	1.24	15	0.4	38	18,898
Kindle reader	0.67	8	0.2	20	10,211
Flahlight	0.50	6	0.2	15	7,620
Telephone	0.40	5	0.1	12	6,096
Heart Pacemaker	0.30	4	0.1	9	4,572
Cell Phone	0.25	3	0.1	8	3,810
Small uProcessor controller	0.18	2	0.1	5	2,743
Hearing aide	0.05	1	0.0	2	762

Fig. 3.5 – EMP induced voltages into Building victims

Do you really know how devastating and ugly the outcome of an EMP incident can be?
If not, read William Forstchen's "One Second After"
Then, be prepared.
Also, see page 9 of this book.

Chapter 4

EMP Protecting a Building without Solar

Chapter 4 Overview

This chapter shows how to provide EMP protection to an entire building, first without solar rooftop addition (which is covered in a later chapter). Remember if the building cannot be protected against EMP, per se, then any other additives are academic since most/all the building's electrical/electronic contents have burned out as discussed earlier. On the other hand, if the building is EMP protected and there exists no solar or other energy backup, the building survival is almost meaningless as there is no electricity either.

Section 3.4 determined the amount and kind of EMP protection needed for an entire building before solar-PV (and in a later Vol. 2 for the electric grid). This chapter illustrates how this protection may be achieved at the component level along with different options for EMP hardening solutions. It further shows that aging effects, such as in building grounding components and earthing, can significantly time compromise EMP protection and how this can be mitigated.

In EMP protecting of the entire building, all outside windows and doors need special shielding treatment. All cable entries must have surge suppressors and possibly filters. The outer shield of entrance cables must be bonded to the building shield and earth grounded to keep the radiated EMP from flowing onto the building shield where compromise otherwise develops. Even the earthing ground (sand or loam soil, wet or drought) significantly impact the shielding performance of the building facade.

The remaining sections address testing to confirm the achievement of the needed 80-dB shielding (below 64 MHz) and maintenance required as the building and its grounding age, together with some measures of cost.

4.1 How Shielding Works

Figure 4.1 is a diagram explaining how shielding works in simple terms. Basically, an EMP or EMI (electromagnetic interference) source at the upper left (marked E and H field) strikes a barrier (usually metal) and reflects back to the left at the same opposite angle. The ratio of the incident electric or magnetic field to the corresponding reflected field is defined as the reflection loss (RL_{dB}) measured in decibels.

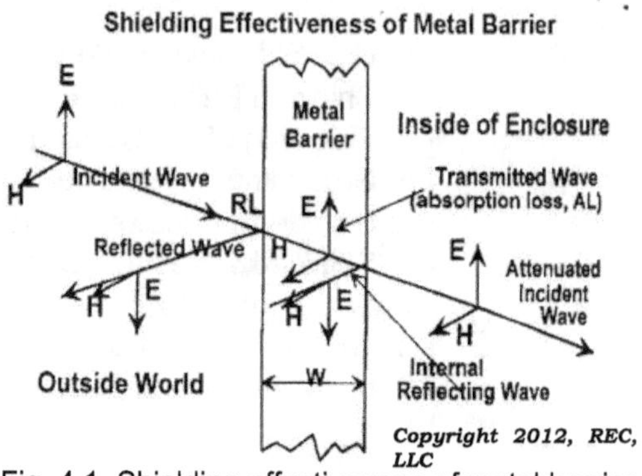

Shielding Effectiveness of Metal Barrier

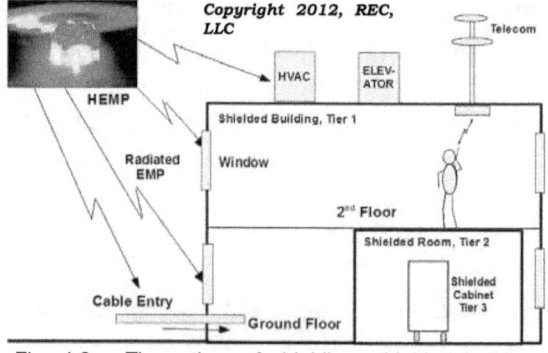

Fig. 4.1–Shielding effectiveness of metal barrier

A second loss, due to absorption (AL_{dB}) of the wave traveling through the metal barrier, is shown inside the metal. Shielding Effectiveness, SE_{dB}, is defined in Eq. (5.1) as a combination of both losses:

$$SE_{dB} = RL_{dB} + AL_{dB} \qquad (4.1)$$

For more details, see the author's handbook on "Grounding and Shielding", published by Interference Control Technology.

4.2 Shielded Building, Rooms and Cabinets (other than the first paragraph, the non-technical reader can skip this section)

Figure 4.2 and Eq.(4.1) illustrate that there exists three possible levels of shielding. The benefit of shielding the entire building to 80 db means that nothing else inside needs to be shielded. So existing climate controls, cell phones, iPhones, iPads, laptop computers, peripherals, and all the rest of electrical and

Fig. 4.2 – Three tiers of shielding: shielded building, shielded room and shielded cabinet or box.

electronic devices in chapter 3 can continue to be used. However, shielding a building to 80 dB in some cases may become expensive (perhaps, for example, more than 10% of the entire building cost by 2014; and a lower percentage later).

Consequently, one or more internal shielded rooms could be added where the more sensitive items would be located, and the entire building shielding can be correspondingly reduced. For example, suppose a cheap aluminum foil or sprayed copper paint on the walls, ceiling and floor (bonded and secured) is used for the inside building conductive skin to reduce its requirements to, say, 40 dB. This may appear to drop the price of the building shielding. But, be aware that the electrical wiring in the outside walls are still not protected; _therefore,_ _this will not work._

However, if the items in the proposed internal shielded enclosure have a radiated susceptibility below 10 V/m (the assumed conditions in chapter 3), the additional shielding can be achieved in an internal screen room or cabinet or box size, as applicable. Shielded enclosures of the quality offered by the EMC manufacturers are shielded to about 120 dB and represent a huge overkill and, therefore, would not be used here.

The shielded building skin is best placed on the outside with sheet metal building walls bonded where joined and each of the six sides of the building (roof, basement floor, and four sides) is bonded along its entire length. Do not try to shield the wall board on the inside as this will not work as stated above because all the wiring trapped between the inside and outside skin will act as a gigantic pick antenna and destroy the shielding performance.

The amount of shielding needed was computed in Sec. 3.4 from the ratio of expected HEMP radiation level (50 kV/m per MIL-STD-188) to the susceptibility level of the victim devices (EU and other compliance levels are typically required to meet 10 V/m as the standard level. So 50 kilovolts per meter/10 volts per meter = 5,000 ratio = $20*\log(5,000) = 74$ dB + 6 dB for safety margin = 80 dB.

Eq.(4.2), next page, gives the simple additive formula in dB for the three levels of shielding shown in Fig. 4.2: (1) outside building skin, (2) inside shielded room, and (3) shielded cabinet, rack or console. Note: 80 dB is the total amount required for HEMP protection. But the illustration indicates that there are three levels of protection possible:

$$\text{Total Shielding, } S_{dB} = \text{Build}_{dB} + \text{Room}_{dB} + \text{Cabinet}_{dB} \qquad (4.2)$$

for building skin, only, $S_{dB} = 80 \text{ dB} + 0 \text{ dB} + 0 \text{ dB} = 80 \text{ dB}$

where: Build_{dB} = building outer skin shielding in dB

Room_{dB} = shielded room (enclosure) in dB

Cabinet_{dB} = shielded cabinet in dB

Fig. (4.2) shows that the windows, rooftop elevator shack, A/C heat exchanger, and all cable shields penetrating the building must not have SE less than 80dB (below 64 MHz) from earlier discussion in Section 3.4.

4.3 Shielding of the Building Facade

How do we shield the building facade (outer skin)? What constitutes the basic material under the building skin or envelope, such as with vinyl siding? In the U.S., it is typically a plywood lath sheathing in the North and concrete blocks, stuccoed over, for residential in the South. For commercial it may be aluminum composite panels, copper or stainless steel sheets, weather-board, etc. These and other materials are dependent on whether or not the building already exists or has not yet been designed or built. Most specifics of this discussion are beyond the scope of this handbook and involve important architectural matters.

A later section of this book will discuss BI-PV (third generation materials = building integrated, photovoltaics) and involve the simultaneous decision of combining solar BI-PV additive which also becomes the building facade. Of course, this is best done for a new building. However, BI-PV would have to be shielded. This forthcoming issue is not discussed in this book except for some topics in Appendix C.

If the facade of the existing building permits the direct addition of an aluminum foil, then the household, 1 mil (= 0.001 inches = 0.0254 mm) or a more ruggedized version (for example, 3 mils) provides all the shielding needed. For example, our shielding effectiveness computer program #330A provides the following shielding performance for 1 mil aluminum foil.

One mil of aluminum foil provides 96 dB of shielding. From Chap 3, 80 dB was determined to be the required amount of building skin shielding to protect against a EMP incident for most applications. So there exists sufficient shielding. But, how are sheets of foil to be bonded to their mounting material siding and how are they mated together at their edges?

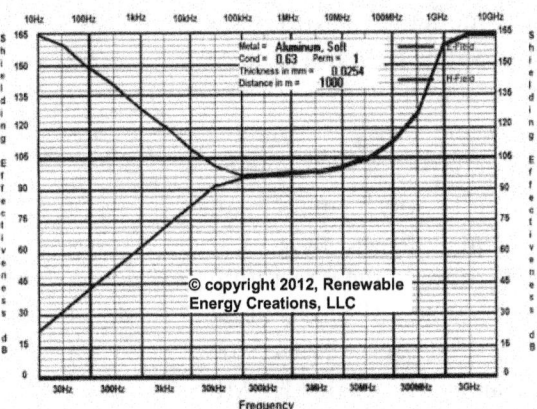

Fig. 4.3 – Shielding Effectiveness in dB of Aluminum Metal foil vs. Frequency

Basically, an adhesive spray is made on the mounting material back and the foil is placed thereon. A squeegee may be used to smooth the mounting. However, the foil overlap should approximate one inch (2.5 cm), and a masking tape used to secure the overlap junction. Do not spray the foil adhesive in the 2.5 cm region as *metal must be bonded to metal* without any other material in between to ensure a high conductivity.

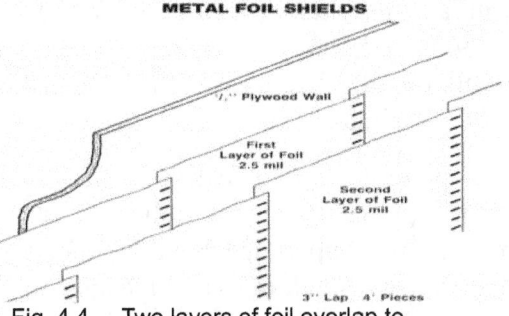

Fig. 4.4 – Two layers of foil overlap to ensure against handling and aging effects

The foil must end at each window sill or outside door sill or frame as the window/door will receive its own shielding. An electrical bonding agent or gasket (described below) is used to electrically connect the building facade with each window and door periphery.

To demonstrate the need for meticulous workmanship, suppose that a hole in the shield foil as small as 0.1 inch (2.54 mm) resulted. What is the new shielding performance of the foil with the hole? Fig 5.5 shows the resulting performance of 58 dB at 64 MHz – the frequency of the second node in the HEMP time-domain pulse discussed earlier in Section 3.4. This hole can be

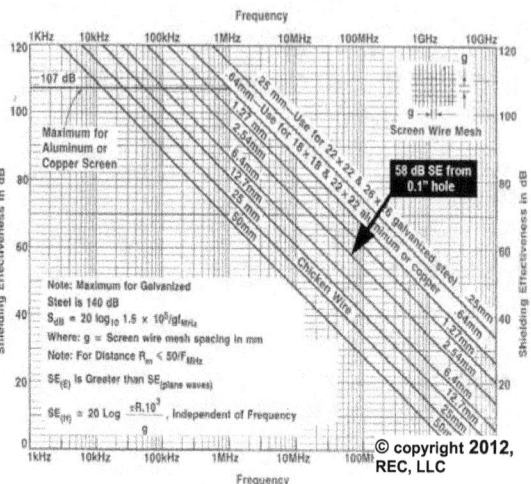

Fig. 4.5 – Shielding leakage with holes

developed in many ways. One way: suppose a workman secured the foil with a screw into a plywood siding (accidentally or intentionally); then he removed the screw for whatever reason! This explains why two layers of foil were used earlier and two layers of wire mesh screen are used on windows and solar panels.

One possible option to the above foil is to use copper or aluminum paint, applied by brush, roller or spray. One source (*LessEMF.com*) reveals a copper latex paint reported to produce less than 0.1 ohm/square (shielding effectiveness of >72 dB below 1 GHz) for 2 mil deposit. Five mil in two coats will produce the required 80 dB at a paint cost of roughly $2 per square foot. ($21/sq.meter). Exclusive of windows, doors and other building skin discontinuities, a 20,000 sq. ft. building will cost $80k in conductive paint (Ed: possibly reduced to ≈ $25k for quantity of over 200 gallons).

A finishing non-conductive protection coat of latex paint is applied. One major benefit of the paint approach is the relatively easy application by spraying and the avoidance of foil overlaps, and electrical gaskets except at windows, doors, etc. described below.

Duralux Aluminum Marine Boat Paint Green, or equivalent may be a viable protection at a significantly lower cost per gallon ($85). In all cases the surfaces must be clean free of dust, dirt, oxides, etc. Sand blasting may be a necessary first step. Application to new buildings is less expensive than retrofits to existing buildings.

Remember, the building has six sides: the four vertical sides, usually $90°$ apart, the horizontal top and bottom. More complex facades (e.g: dormers and gutters) have more faces; the same physics and engineering requirements still apply, but realizability becomes somewhat more complex. In any event, the remaining shielding involves bonding the adjoining meeting sides with the proper material.

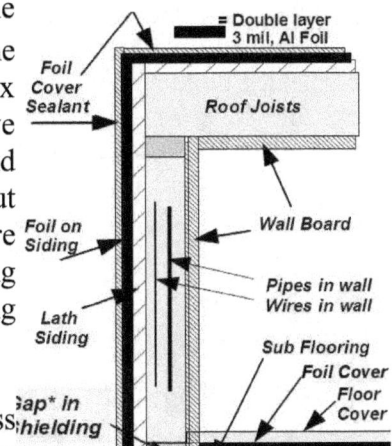

Fig. 4.6 – Detail showing shielding gap before closing in existing retrofit

Of the six sides, the only bonding access problem is in *retrofitting* (not new construction) the bottom of the lowest floor of existing buildings. As shown in Fig. 5.6, there exists a gap between a foil or other conductive flooring on the inside bottom and the four vertical sides of the remaining shield faces on the outside.

How to close this gap? There are different methods, but they go beyond the level of this book. Of course, to a new building just being planned or being built, this is not an issue as the bottom floor shield is extended beyond outer frame to be bonded to side shielding (Fig. 5.6).

Regarding electrical gaskets used to seal the adjoining metal sides, Fig. 4.7 shows options available from Tech-Etch for the engineering community. The figure also shows other electrical gaskets useful in bonding other types of mating

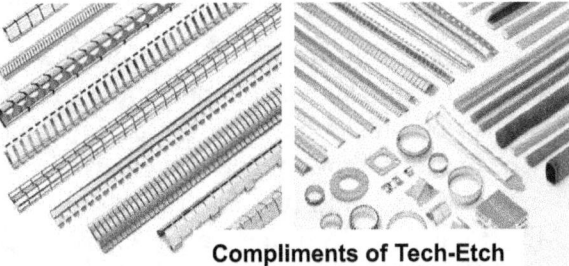

Compliments of Tech-Etch

Fig. 4.7 – Many different electrical gaskets available for selection of the correct type for any particular application. Right central wire mesh type for window sills.

electrical parts, discussed elsewhere in this handbook.

4.4 Grounding of Shielded Building (for technical readers)

Fig. 4.8 shows the shielded building including building skin, shielded windows, shielded HVAC and elevator shacks and other external elements. The shielded building could float in space without grounding and the shield works fine. No grounding is required. However, there are other concerns such as if the shield becomes charged up to any voltage above ground due to whatever stray radiations (or accidental connection of a hot AC lead to building skin), it constitutes a shock hazard to entering and exiting people. Also, for lightning control. That's why the building skin must be grounded (earthed).

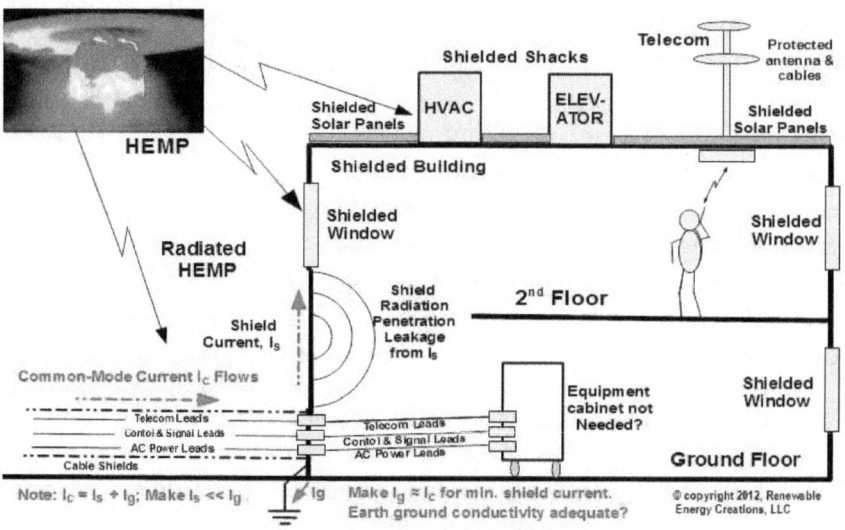

Fig. 4.8 – Shielded building with grounding compromises

Also, as shown at the lower left, Fig. 4.8 telecom leads, control and signal leads, and AC power leads must enter the building. These cables must also be shielded. Since the cable shield and the building shield may be at a different voltage, the entering cable shield must be bonded to the building shield (usually a metal plate. The object is to divert the EMI (induced electromagnetic interference shield currents) and grounded to earth, where they are harmlessly dissipated. But, the devil is in the details.

The sketch of Fig. 4.8 illustrates the situation repeated thousands of times throughout a state or a country of a developed nation. Radiated EMI (or HEMP) is coupled to a cable shield where EMI surface, common-mode currents, I_c, are grounded at the shielded building entrance (assuming a best case in which the building is shielded). It is typically assumed that the EMI currents will flow down the ground rod(s) (see I_g) and dissipate harmlessly into the absorbing earth. But, some EMI currents (I_s) still flow on the building shield (that skin current is capacitive coupled back to ground to close the loop). Those currents may then penetrate into the building skin where they can upset or burnout internal victims. So, how successful (e.g., how low?) is the low-impedance grounding or earth system and how successful is the building shield working? It all gets down to the quality of each and attention to pertinent details.

Remember, the only shield current that should flow is from the terminated, direct radiated EMI onto the building shield, per se. It should not contain additional currents that come from any cable entry ground which can exceed the terminated EMI surface current on the building by large amounts. The degree to which it is exceeded is a direct measure of the reduction loss in building shielding effectiveness.

Fargo Clamp

Compliments of ElecDirect

Fig. 4.9 – Grounding wire, rod and pipe clamps

If the building were built on sand or some non-moisture holding soil, or a drought existed in the area, then the grounding system will not work as well (or poorly). Or, if the ground Fargo clamps (Fig. 4.9) eroded out or became disconnected, then what? (Note Fig. 1.1, *Measurement of Soil Resistivity*, in MIL-HDBK-419 speaks in detail to the number of ground rods, ground wells and earth electrode subsystem to achieve the grounding objectives).

Experts, like Dr. Radasky and the authors of this book, have sadly witnessed over the years a plethora of meaningless grounded installations on multimillion dollar infrastructures and equipment. How often is a grounding system inspected and/or tested for numerical compliance? Probably not at all once it is installed for the first time. So, then are all the millions of cell-phone towers, TV, radio, telecommunication and other towers protected? Certainly, many or most are compromised. Does the insurance carrier ever inspect? Never or rarely, unless there is a fire or intrusion damage. So, unbeknown to many, an electrical or electronic system installation grounding system for shock safety and lightning control, may, in reality, be an injury or a death trap waiting to happen.

Can you testify if your own facilities or other installations are properly grounded? What criteria or standard do you use? Who signed off on the inspection? Still "grounded" five years later?

(Note from Donald White: Here is a short-short story you may enjoy. While my visiting UK friends were at Jefferson's home in Charlottesville, Virginia, I departed from the line of waiting patrons, to examine the lightning protection system of the large majestic old oak trees. The lightning rod stood above the trees and the #6 AWG down conductor was in place. But, the grounding Fargo clamp no longer was connected to the line – perhaps separated by a lawn mower hit. The oxidation of the two suggested that this may have happened years ago. What does this tell you?)

As it develops the entire discussion may be academic from an EMP point-of-view since few non-government buildings are EMP (shielded) protected in the first place today. For those situations, all the interior electronics are fried in an EMP event – grounded or not.

EMP, lightning and EMI won't go away. Grounding is an important part of the protection system, but shielding and filtering are of equal concern. Many of the details of shielding and filtering are too technical for a book intended for non-technical readers.

Good Electromagnetic Grounds

Because RF (radio frequency) current flows on the outside conductor surface only at MHz and GHz frequencies, circular conductors are the poorest configurations (their cross sectional periphery-to-area ratio is the lowest). As Shown in Fig. 4.10, contrast this with a strap, cross section or better yet, a thin foil which

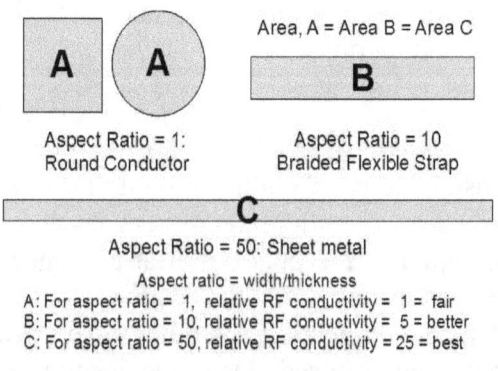

Fig. 4.10 – Performance of Grounding Conductors

conveys the EMI from the shield bond down to the earthing conductor rods more effectively.

If the soil has poor conductivity (sand, drought), it can be increased in the region by adding bauxite, bentonite or a proprietary grounding material. The answer is to make the I_g/I_s current ratio in Fig. 4.8 as high as possible or meet a specified numerical objective for a specified installation.

Numbers are not used here as they will bore most non-technical readers and have little meaning. But, the name of the game is in the numbers; their cost and performance are all a matter of achieving numbers, which presumably have a supporting rationale.

4.5-Shielding Windows, Doors & other Building Skin Leakage.

Windows cannot employ electro-chrom glass as this blocks light with little shielding, nor can one have a few microns (very thin film) of vapor-deposited metal on the window as a shield since one could not then see through the window. Although a one micron (very thin film) of silver deposited on glass offers an RF shielding effectiveness of 80 dB, it also blocks the optical transmission.

Fig. 4.11 – Wire mesh, shielded windows

Therefore, a metal screen mesh (Fig. 4.11) is considered for covering the windows which passes almost all the light (optically) but blocks the HEMP radiated transmissions by 80 dB below 12 MHz (depends on the mesh screen spacings). The required screen mesh separation distance, d, to achieve this shielding, is calculated by:

$$SE_{dB} = 20Log_{10}(150/(d*f_{MHz}) = 80 \text{ dB} \qquad (4.3)$$

where: f_{MHz} = 12 MHz from the geometric mean of the first and second corner frequencies in Fig 3.2, Section 3.4)

Solving Eq. (4.3) for mesh separation, d, produces a distance of d = 1.29 mm. This corresponds to 25.4/1.29 or 20 wire mesh openings per inch (OPI). This OPI is mentioned since mesh shielding is usually reported in OPI by manufacturers, and is available up to 100 OPI.

Do not use higher OPI as the light transmissivity begins to suffer and this will adversely affect the solar panel efficiency of converting sunlight into electricity.

To facilitate using Eq. (4.3), it has been graphed earlier in Fig. 4.5, One word of caution. The manufacturers seem to show a shielding limit of about 60 dB. This is partly because of the use of wire mesh and the perimeter of the shield is not bonded with a low enough conductive gasket. This is demonstrated by a copper screen shielded room which shows a lower limit more nearly 85 dB. While this topic is beyond the intentional technical limits of this handbook, Fig. 4.12 shows proof of this remark.

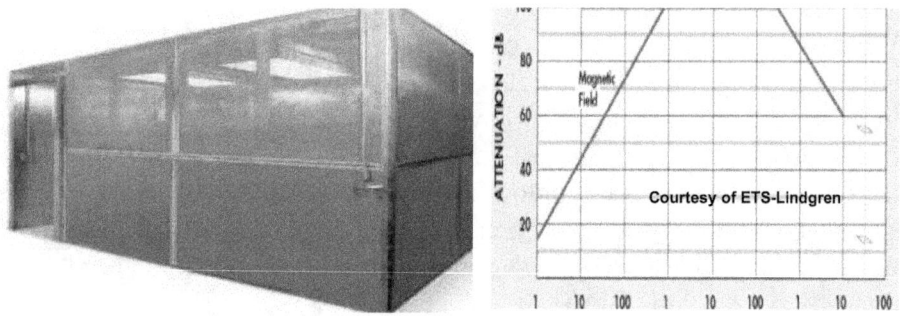

Fig. 4.12 – Shielded screen room with more than 85-dB shielding below 1 GHz

Shielded External Doors

One or more doors permit entry into and exit from the shielded building. For each such door, a vestibule must be generated such that a set of two doors is used with interlocks so that not more than one door can be opened at any time. This will avoid a fortuitous situation in which both doors are temporarily open at the moment an EMP incident occurs, which would otherwise compromise the entire integrity of the building shielding effectiveness.

The vestibule can be added either external or internal to the existing outer door, whichever is the more practical, available room, cost, etc. Whether external or internal, the vestibule must also be shielded on all six sides to preserve the building's shielding integrity

The doors may be shielded in any of several ways as already discussed for the building shielding. Bonded aluminum (or other metal) foil on either or both door faces covering all door five or six sides. Around the periphery of the door is bonded shielded finger stock, wire mesh or other suitable bond (see Fig. 4.7). Finger stock is often used on door seams because its life expectancy goes into the thousands of times of openings and closures. The stock

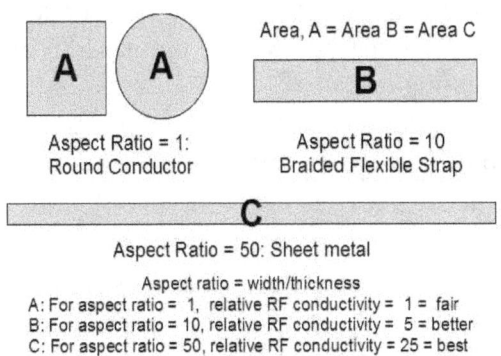

is mated with a bladed seam so that no contact can accidentally be made with clothing of entering/exiting persons.

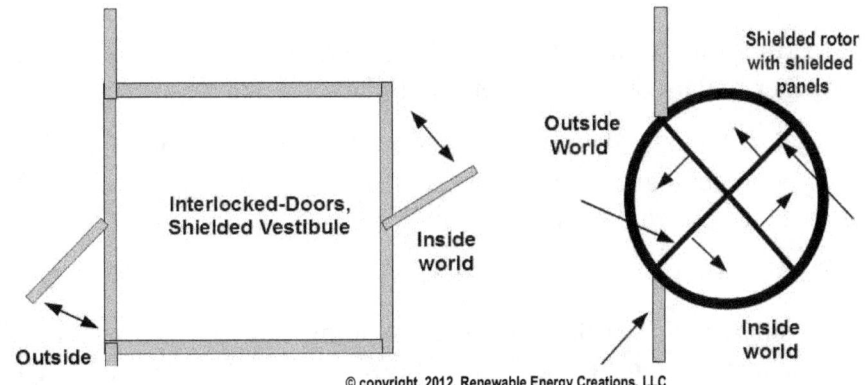

© copyright, 2012, Renewable Energy Creations, LLC

Fig. 4.13: EMP shielded building entrance/exit options.

The finger stock or mesh will have to be periodically replaced if the doors are opened hundreds of times per day. Therefore, a maintenance cycle must be established.

Fig. 4.13 also shows another option to construct a rotating shielded entrance frequently seen on commercial office buildings. Here, of course, the entire assembly is shielded including the four rotor partitions.

4.6-Shielded Wires and Cable Entries

As shown in earlier Fig. 4.8, three cable and wire classes enter nearly every building: (1) telecom (telephone, Ethernet, and other hard-wire communications) leads, (2) control and signal leads for controlling certain activities and reporting back elsewhere with status data, and (3) AC power mains to supply 120 VAC, 240VAC or other voltage to run the building's many operational loads previously discussed. Unless the cables are already placed in conduit, buried and/or shielded, their shields must provide the required 80 dB below 64 MHz isolation to the entering leads.

The interconnecting wiring between and among all solar panels, inverter (if micro-inverters are not used) and the down conductors to the service entrance panels must also be shielded to 80 dB. A solid thin wall copper tubing can also be used as the shield. However, to ensure the benefits of flexibility, knitted-wire mesh shielding

Courtesy of
Tech-Etch

Fig. 4.14– Wire mesh, cable shields

can be used on all cabling instead. Fig. 4.14 shows a compliant shield for interconnect wiring. All shielding must have an overcoat of weather sealant or outer jacket if not placed inside a conduit.

As explained above, the cable shields are bonded to the building shield usually at a metal grounding plate welded to the building shield at cable entrance (see above discussion on building grounding).

4.7 Surge Suppressors and Filters

Do not suppose that each of the three building cable types contain EMI (electromagnetic interference) or some residual EMP protection. This, of course, depends on several variables such as cable type, buried above or below ground, inside a metal conduit, length of cable before building entrance, etc.

The telecom cables require a surge suppressor rated to clip all surge voltages above some minimum, such as 100 volts (well above the intentional levels), that may result from a residual EMP. The control and signal leads are confronted with a similar situations. Filters are not used as they may not cut off at high enough frequency unless all future data rates are known well in advance.

AC power mains is another matter since it is supposed to contain only 120 VDC, 240 VAC or other service voltage values. The cutoff frequency should not be just above 60 Hz (or other power line frequency) since the inductors and capacitors become big, bulky, weighty and expensive. Therefore, most have a low-pass cutoff frequency at about 1-10 kHz.

To protect insulation burnout of the filters with a EMP incident, they are preceded by a surge suppressor rated at one of the above voltages. This depends on whether the AC service is furnished underground or above ground, enclosed in its own cable shield, length of cable runs, if the feeding distribution transformer is protected, etc. The next chapter gives a few examples.

4.8-Testing Shielded Building Performance. (for technical readers)

While testing details are beyond the scope of this handbook, in concept it is quite simple. A small van or pickup truck at the left in Fig. 4.15 contains a scanning oscillator or sweeper which feeds a power amplifier which drives radiating test antennas pointing at the building to the right. Inside the building is a tracking receiver driven by matching pickup antennas.

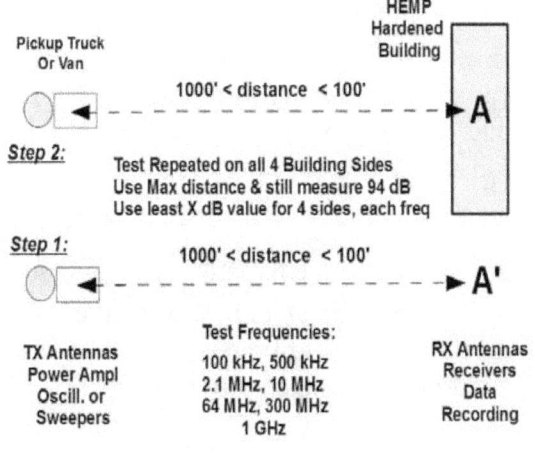

Fig. 4.15 – Test setup for determining shielding performance

The test configuration is first calibrated as a reference with both transmitter and receiver configurations on the outside (step 1). Tests are made at the seven frequencies indicated in the drawing, and the attenuator settings recorded. Then, maintaining the same distance between transmitter and receiver, both are moved as shown in step 2 with the receiver configuration moved to the middle of the building and attenuator settings again recorded to get the same levels. The difference between the attenuator settings in step 1 and step 2 constitutes the building shielding.

The tests are repeated for the other remaining three sides of the building, recording the shielding effectiveness of each side for all seven frequencies. Then for each frequency, the corresponding four side shielding effectiveness are compared and the smallest value selected. These results constitute the building EMP shielding effectiveness. They are compared with Fig 3.4 to determine compliance.

There is a more automated procedure than just explained. It is discussed in chapter 6. Also, if the building is large (several floors or acres in size), the transmitter may be located in a helicopter or blimp.

4.9-Maintenance Considerations

Everything ages including all buildings and their contents. For the outside, the more severe the climate and weather, the more rapid degradation and variance in performance is expected. In addition to shock safety, building grounding will determine how well the building EMP shielding has held up. As stated earlier, the soil loaminess (somewhat dark, and good for growing grass or other vegetation) makes for good grounding as it holds moisture. Arid soil like sand is very poor and low on conductivity. It may need to be impregnated with bauxite (an aluminum base particulate) or proprietary grounding compounds to permit better grounding and earthing into a much lower impedance.

As observed, testing for grounding effectiveness will show up in the above open air shielding effectiveness tests. So a building's earth and grounding should be tested by a licensed and certificated technician to add to the insurance protection.

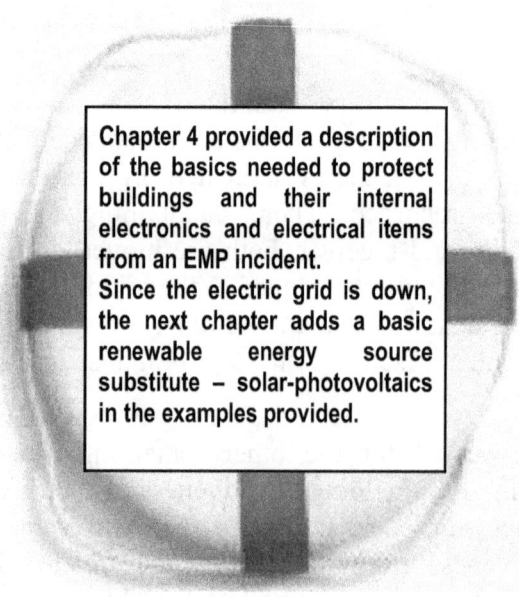

Chapter 4 provided a description of the basics needed to protect buildings and their internal electronics and electrical items from an EMP incident.
Since the electric grid is down, the next chapter adds a basic renewable energy source substitute – solar-photovoltaics in the examples provided.

Chapter 5

Adding Building Solar Rooftop

Chapter 5 Overview

This is the first chapter that solar-PV is being discussed in length. Solar is "green" (no carbon pollutions) energy and in 2012 infrequently provides electricity at a price less than that provided by the electric utilities. (Solar electricity is expected to be on par with fossil fuel electricity by 2017-2020). The necessity here is that solar (or another renewable energy) provides the electricity after a EMP incident when the electric utility can no longer do so. Repeating for emphasis: This chapter speaks to typical solar rooftop installations *before* solar EMP hardening. Chapter 6 addresses the EMP protection required for solar and how it is done.

Chapter 5 also discusses the three generations of solar (crystalline silicon; thin films; and solar paints, inks and dyes). Then "How solar works" is reviewed.

The rest of the chapter speaks to the solar component victims not addressed in the preceding chapter since they will have to be EMP protected as addressed in Chapter 6. Meanwhile, the solar components include solar panels (and solar panel mounting racks), micro-inverters and inverters, new wiring added and a solar smart meter.

The chapter closes with a section on solar cost expectations, break even and ROI from 2012 up until 2020.

Since this chapter is a continuation of the previous chapter, solar-PV rooftop victims have been added to the non-solar building, So to the previous building victims, we have added solar victims as shown in black, Fig. 4.1, rightmost column.

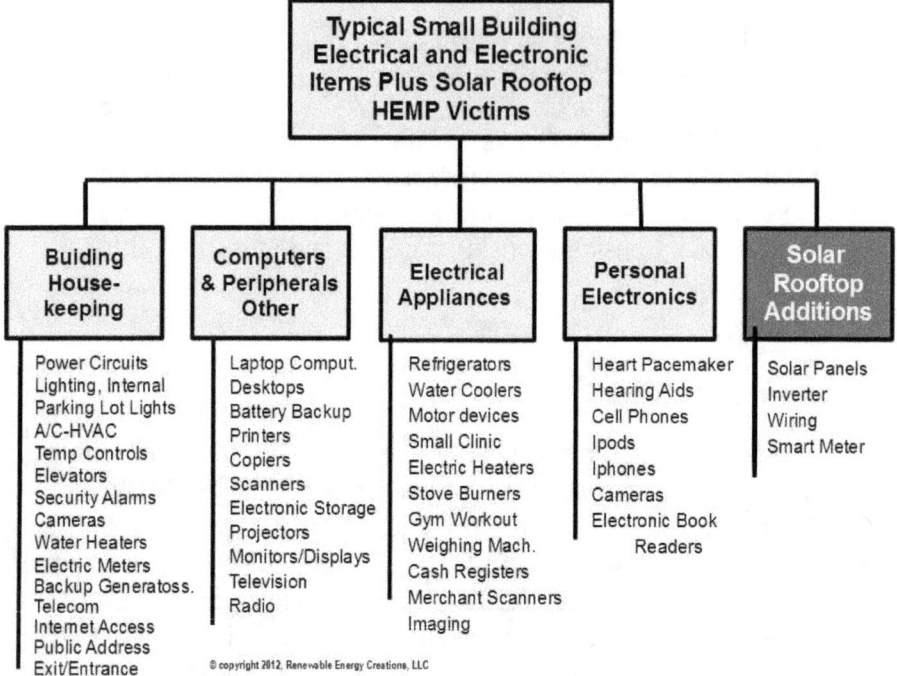

Fig. 5.1 – Solar rooftop EMP victims added to small building victims

These solar-rooftops added to subject buildings provide the power to the buildings to reduce the electric power company charges, perhaps remove the charges altogether provided sufficient solar panels are installed. However, they need the utility power at night and overcast days. Solar is also needed in the event of an EMP incident as there is then no surviving electric utility power for the building. If the new solar installation follows the 80-dB shielding guidelines of previous chapters, then the installation becomes nearly self sufficient, provided additional battery and/or engine generator is added. (See Chapter 7 for details.)

Figure. 5.1 shows that the new susceptible victims before EMP protection are solar panels, micro-inverters or larger powered inverters, interconnected wiring and the added smart meters. Their specifics for EMP protection are discussed in chapter 6.

5.1 Solar-PV Generations and Performance

There exists three generations of solar-PV technology:

1st Generation, Crystalline Silicon solar represents the most popular solar-PV with about 85% of the solar installations as of 2010. It is the most expensive and about 20% efficient in converting sunlight into electricity. It is the oldest of the solar technologies and has a 25 year guarantee and a 40+ year life expectancy for most locations.

2nd Generation, Thin-Film solar is made from amorphous silicon (the least favorable of the 2nd generation), or from popular cadmium telluride (CdTe) or copper, indium, gallium, selenide (CIGS) cells. In thick-

Table 5.2 – Solar rooftop technology and median installed price

Solar Generation & Technology		2010-2011 Efficiency*	2010-2011 Price $/Watt	2013-2015 Efficiency*	2013-2015 Price $/Watt
1st Generation	Crystalline				
	Silicon	19	$5.00	20	$3.00
2nd Generation					
Thin Film	Amorphous Si				
	Cad Telluride	11	$2.50	13	$1.50
	CIGS	10		8	
3rd Generation					
Inks and Dyes	Rooftops			5	$0.30
	Siding			2	

ness it is less than a human hair in diameter, but mostly formed into rigid, glass covered panels. Efficiencies are reported at about 11%, but First Solar Inc. reported that they can manufacture new ones with a 14.4% efficiency in early 2012. You should beware; 2nd generation efficiency values are contentious, as they are based on a different measure of efficiency than first generation. This is discussed below.

3rd Generation, Paints, Inks and Dyes, still, mostly in experimental stages and in a few years expected to become practical, competitive, and inexpensive installations. While applied vertically along the outer skin of a building, their horizontal pointing inefficiency is low (equivalent efficiency about 5% at best) but their cost is very low. So the resulting yield/cost ratio is becoming favorable. Because the 3rd generation solar is still in its early stages, it is not discussed further here. Instead, you can view Appendix C for further information.

More on Solar Cell Generations, and Efficiency

Many, if not most who are knowledgeable of solar cell technology, may say that first generation silicon solar is more efficient, reliable, and has a longer life expectancy than second generation, thin film, solar technology. They further argue that silicon solar technology has been around for 50 years, used by NASA in space, and has a large reliable data base, while thin-film has only been out for about 15 years. The confusion or problem here is that the *measurement criteria* and other metrics of performance *are different* for the two solar generations. And, there is always the "turf politics" problem of NIH (not invented here).

Thin-film solar companies are often weary of being asked about their conversion efficiency, which is basically defined for the first generation as the amount of sunlight a solar panel can convert into *peak* power. Part of the problem is because the thin-film manufacturers say the efficiency standard is flawed. Therefore, some thin-film companies are pushing for a new and more meaningful or relevant standard.

The measurement criteria for first generation solar is based on questionable assumptions. For example, peak power output (not average or under conditions representative of a real-world installation) is used for silicon. Second generation solar (for example, CdTe and CIGS) argue that the total kWh energy output per kW of power installation over a unit of time is the real-world use and test. Most first generation advocates say "no" (because this will make their product look less favorable and even more expensive).

Another reason for discord is that thin-film solar can make electricity in diffuse light (for example, overcast skies), while regular silicon-based panels need more direct light. Thin-film advocates say this means thin films can produce more electricity throughout the day if graded by not using the silicone standard of measurement.

The discussions here are not based on a contentious efficiency term based on peak power output, but rather the kWh/kW/day approach. (kWh/kW/day = kilowatt-hours of energy per kilowatts of installed power per 24-hour day). When this is used, thin film becomes more "comparable" to silicon daylight conversion, even though its silicon definition places it at about 65% "efficient" relative to silicon. Unless stated otherwise in this book, the default efficiency of thin film is about 65% of that for silicon, based on energy daylight consumption – not peak power.

In summary it is absolutely essential that the matter of silicon peak power efficiency vs. thin-film kWh/kW/day be resolved ASAP. Otherwise we are dealing with apples and oranges, and saying that they are fruit resolves nothing, but only exacerbates and prolongs the problem and communication misunderstandings.

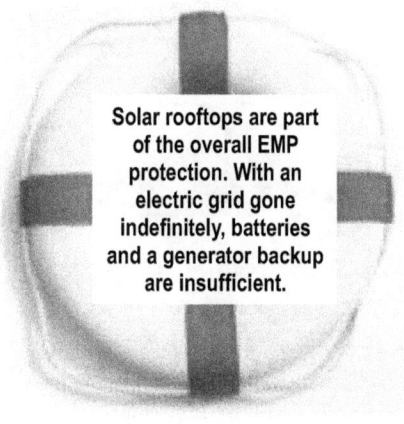

Solar rooftops are part of the overall EMP protection. With an electric grid gone indefinitely, batteries and a generator backup are insufficient.

5.2-How Solar Rooftop Installations Work

The accompanying illustration shows the six principal parts of a solar rooftop installation for a residential application.

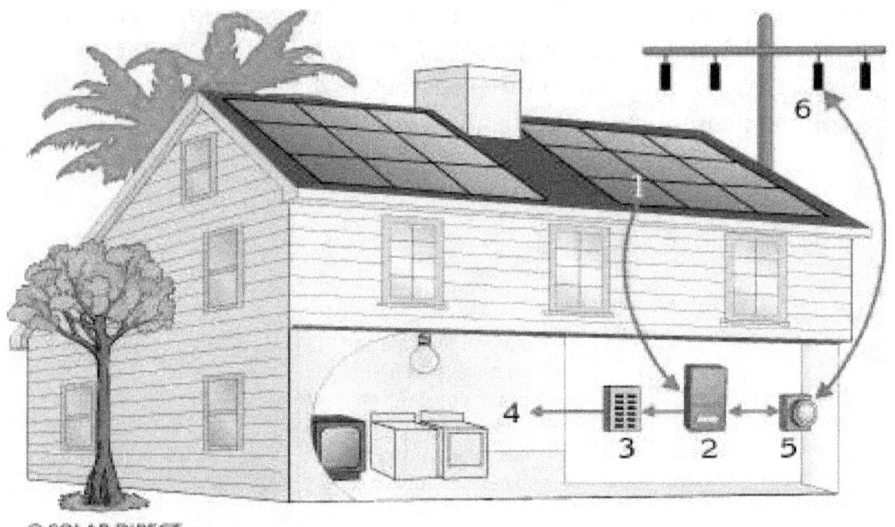

Courtesy of **Solar Direct**
Fig. 5.3 – Solar rooftop system installation showing six major parts

(1)- Solar-PV (photovoltaic) modules convert sunlight (or diffused solar) into electricity by the photo-electric effect. The PV modules, generally organized into solar panels, generate low voltage DC (direct current) electricity. Their output is connected to an inverter to produce (convert to) AC Voltage.

As explained in Section 5.1, a PV cell is most often made of a thin wafer of silicon modified with small amounts of other materials that give the silicon wafer special electrical properties. When sunlight hits a PV cell, it produces an electric current. PV cells are connected together in a solar module, which has a non-reflective glass front, a protective insulating back sheet, and aluminum frame for strength and mounting.

(2)- The inverter transfers the 12-48 volt solar DC power output into 120-240 VAC (alternating current) electricity for household consumption. The inverter may also provide safety functions such as automatic shutdown of the solar electric system in the event of a power failure.

(3)- Existing electrical panel, containing circuit breakers for each circuit between 10 and 50 amps. The panel gets its input from either the power company (electric grid) or the AC output from the solar inverter, depending on which is connected.

(4)- Electrical loads, consisting mostly of appliances (air conditioning, water heater, clothes dryer, oven, stove-top burners), electric hand tools, and electric lighting.

For systems with a battery backup (optional, but usually located in areas where the electric grid does not exist), the inverter also regulates the charging of batteries. The electricity stored in the batteries can be used at night, or during overcasts days or electric blackouts. The battery output is connected to an inverter, similar to the solar cell output.

(5) A valuable feature of photo-voltaic systems is the ability to connect with the existing utility electric power grid which allows solar owners to sell unused electricity back to the utility with a plan known as "Net Metering."

(6) At times when the electricity produced is not being used, the meter will run backwards - selling the electricity back to the utility power grid at retail rates, usually that which they charge the consumer (sometimes lower).

For smart metering installations, the entire system is linked to a Performance-Monitoring Reporting System (PMRS). The PMRS measures and monitors all electricity generated by the system, performs system diagnostics and sends alerts regarding system or equipment issues. The PMRS also tracks weather and generates reports that link weather conditions to kilowatt-hour output.

A few examples of Solar Rooftop Installations

A few examples of solar rooftop installations are shown in the six photos below in Fig 5.4:

A: 12 kW solar panel installation on an A-frame home
B: Tile roof made with bonded thin-film solar
C: Residential apartments with asphalt-solar shingles
D: Warehouse roof containing silicon solar panels
E: Five-acre industrial building with solar panels on roof
F: Commercial office building solar rooftop in downtown area

Fig. 5.4 – Six solar rooftops of residential, commercial and industrial installations

In contrast to residential, 62.5% of the electricity developed in the USA goes for commercial and industrial use. Fig. 4.5 illustrates the many uses and users of electricity in America. Commercial use shows the wide span of building types and uses.

So we may conclude that commercial applications will be the dominant driver of the feasibility and the financial

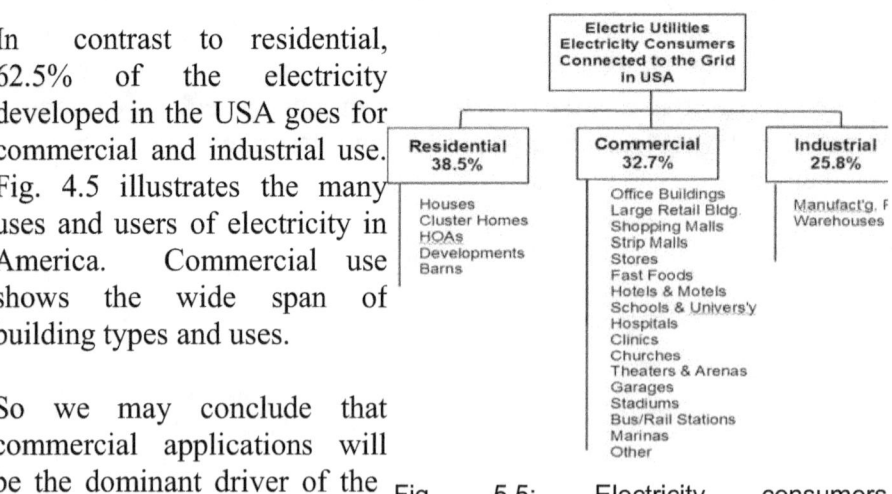

Fig. 5.5: Electricity consumers identification

viability of EMP hardening as well as solar o rooftop power – at least in the formative years.

5.3 Solar Site Visit and Analysis

This first step is usually carried out before a proposal is submitted to a solar prospect. This is important, among other things, that there is no site blockage, Fig 5.6, for the rotating sun by trees or buildings from 20° above the horizon up to the zenith (90° straight up) and back down to 20° before sunset. Otherwise, lost solar energy will result and compensation must be made by taking other measures, including in the price quote submission.

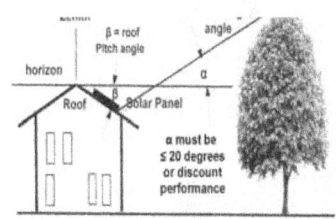

Fig. 5.6
Solar Obstacle Restrictions

Be sure that a sector between the two paths from summer solstice (21 June) when the sun rises most northerly in the Northern hemisphere to winter solstice (21 Dec) when the sun rises 47° more to the South (Fig. 4.7) have been included in site obstruction checks.

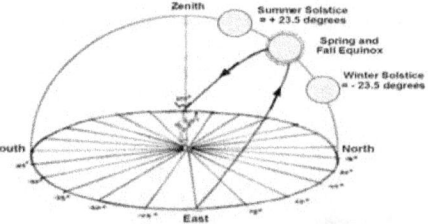

Fig. 5.7 – Solar-Earth Seasons

5.4 Solar-PV Panel Racking

At construction time solar panel *racks* are among the first things to erect in place since they secure the solar panels which cost about 50% of the entire contract installation cost. The mounting integrity is especially important here because the installation, if later EMP protected, must be shielded by 80 dB for an EMP incident protection. No small task! This situation is discussed later in the next chapter on solar panel EMP protection.

Fig. 5.8 – Examples of Panel Mounting Racks

5.5 Solar-PV Panels

Solar panels represent roughly half of the solar installation cost. They are sealed, low-reflecting glass housings containing a number of interconnected silicon (or thin film) solar modules that typically produce an output of about 200 watts ± 20%. The corresponding size is roughly 3 feet x 5 feet (91 cm x 157 cm). Other sizes are available. The frame and backing are made of aluminum and are usually mounted onto the solar racks.

Courtesy, 123rf.com

Some Typical 2012 Solar Panels of Leading Manufacturers

Company ID	Sku	Watts	Efficien-cy, %	Area Sq. Ft.	Cost US$	$$/Watt US$	Watts/ Sq. Ft	Note
Abound Solar	AB1-65	65	9	7.7		Low?	8.4	D
Astro Energy	AE-130 Mono	230			285	1.24		E
Canadian Solar	CSSA-200M	200	15.7					A
Canadian Solar	CSSP-220	220		18.1	596	2.71	11.2	C
General Electric	GEPVp-210	210	14.4	15.7			13.4	D
Kyocera	Kyocera 135	135			306	2.27		E
Sanyo Electric	HIT-220AQ1	220	19		715			B
Sharp	ND-235QCJ	235			348	1.48		E
Solar World	SW-220-Poly	220			340	1.55		E
SunPower	SPR-200-WHT-U	200	16.1					A
SunTech Power	PLUTO220-Ada	200	15.7					A
SunTech Power	STP210-18/Ub1	210		16.5	565	2.69	11.5	C
Yingki Solar	YL165P-23b	165	12.7	14.0			11.8	D

Source: A: http://sroeco.com/solar/most-efficient-solar-panels
B: http://www.solarplaza.com/top10-crystalline-module
C: http://sroeco.com/solar/learn-solar/solar-panel-comparison/
D: http://www.solardesigntool.com/compare-solar-panels-modules.html
E: http://www.wholesalesolar.com/solar-panels.html

Fig. 5.9 – Example of solar panels together with a table of representative variable data of some of the leading global solar panel manufacturers

For second generation, thin film silicon, CdTe or GiGS modules, solar panels are flexible and sealed in their own non-rigid housings. Average efficiencies of CdTe runs about 11% in contrast to roughly 16% for silicon. But, their weight is considerably less than silicon and the price per watt is lower.

Although the table in Fig. 5.9 is not especially detailed, it does contain information useful for applications elsewhere in this handbook. For example, if an average 2,300 sq. ft. home requires 5 kW of power, the above table suggests that 5kW/210W/panel = 24 panels are required @ $500/panel = $12,000. If solar is roughly 50% of the installation cost, the contract price may approximate $24,000 before application of any state rebates and Federal tax credits. See cost considerations discussed later in Section 5.9

5.6- *Solar Micro-inverter and Inverters*

The micro-inverter is a small DC-AC inverter which accepts a typical 12 VDC output from a solar panel and converts it to 120 VAC or other user load value. As appropriate, they are connected in series and/or parallel to produce the desired load voltage. A small micro-converter may be built into the solar panel near its output connector (this is done at the time of solar panel assembly back at the factory) or more frequently mounted separately outside but next to the panel as shown in Fig. 5.10.

Micro-inverters have several advantages over conventional larger central inverters. Even small amounts of shading, debris or snow lines in any one solar panel, or a panel failure, does not disproportionately reduce the output of an entire array. Each micro-inverter obtains optimum power by performing maximum power point tracking for its connected panel.

Fig. 5.10: Solar Micro-inverter

Their primary disadvantages are that they have a higher equipment initial cost per peak watt than the equivalent power in a central inverter, and are normally located near the panel, where they may be harder to maintain. These issues are, however, surpassed by micro-inverters having higher durability and simplicity of initial installation.

When the total rooftop solar load exceeds roughly 10 kW, micro-inverters are usually not used and the inverter takes over. Fig. 5.11 shows three inverter sizes ranging from 10 kW to 500 kW.

Fig. 5.11 – Solar Inverters

5.7 Electric Meters and Smart Meters

Figure 5.12 shows two typical, well-known electric meters. Unless or until they are replaced later by smart meters, nothing needs to be done here. However, the next chapter addresses how they get shielded to protect against the EMP threat, unless their

Fig. 5.12 – Typical Utility Electric Meters

replacement is already shielded to 80 db and their input leads surge suppressed.

5.8-Interconnecting Wiring

Nearly all the AC wiring from the electric utility transformer serving the customer load is unshielded. Unless coming from an exposed utility pole transformer, as shown in Fig. 5.13, the electric wiring is routed down the pole to the ground transformer sitting on a concrete

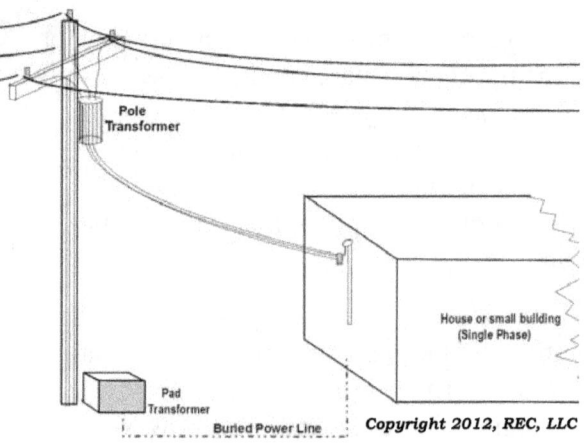

Fig. 5.13 Typical Small Customer Service Entrance

90

pad and the wiring to the customer is buried directly in a trench or placed inside a plastic pipe in the trench. Then, to the service entrance all of which has no shielding in a civil world. Remember, all this will be treated in the next chapter for EMP protection.

The customer premises are also served by cable TV, telephone or one or more of its derivatives and, for the commercial and industrial world, served by control and signal leads derived from other sources. Remember, all these wiring and cabling act like a huge radiation pickup antenna and will become especially vulnerable in a EMP incident unless measures are taken for their protection.

5.9- Cost Considerations, Break Even and ROI

The objective here is to take a rather complicated math computation of break even and ROI for a solar rooftop installation and to convert it to a few relatively simple steps that the reader can readily do for himself in a few minutes after learning the procedure. Parenthetically, we have a computer program which does all this and allows "What-if" war gaming to optimize the return.

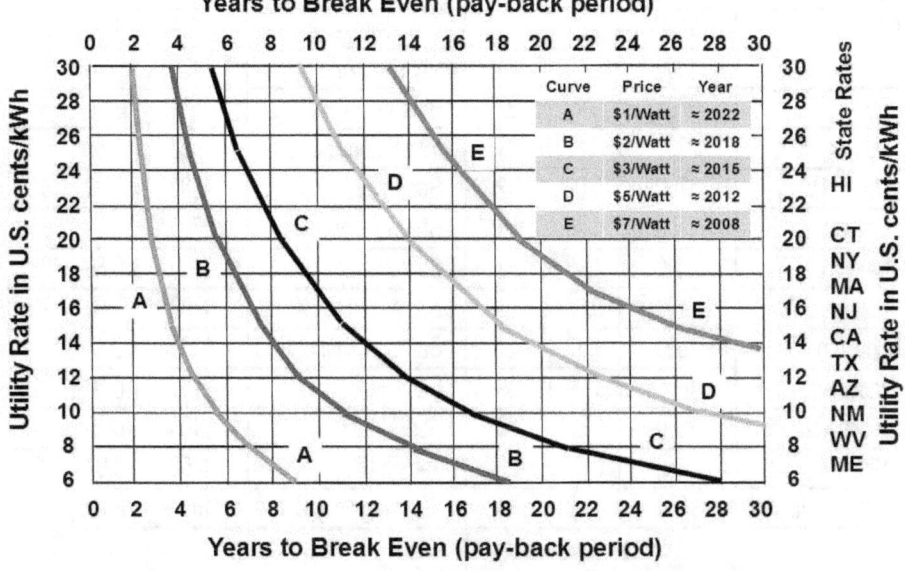

Fig. 5.14 Years for solar installation to break even vs. several variables.

The graph, Fig. 5.14, above depicts the years to break even (X-axis = recover your Investment) vs. electric utility electricity rates (Y-Axis) for solar rooftop installations, costing $1/Watt, $2/Watt, $3/Watt, $5/Watt

91

and $7/Watt with no rebates or tax credits (which can also be factored in).

Assumptions include average five hours of sunlight/day. This is averaged over the seasons, in which there is a 47° change in sunrise/ sunset from summer solstice (June 21 in the northern hemisphere) to – winter solstice (Dec. 21) as previously shown in Fig. 4.7. Note, nothing is said about, the installation site latitude, the compass direction of the solar roof selected and the roof slope angle as well as the photovoltaic performance in different latitude and climate locations. These impact how close/far the solar panels are pointing at the sun in the declination axis (North-South direction).

Fig. 5.15 – City Solar-PV Cumulative Intensity and City Latitudes.

City and State	Intensity kWh/m^2/Yr	Multi- Plier	Latit. degr	City and State	Intensity kWh/m^2/Yr	Mult- plier	Latit. degr
Atlanta, GA	1800	1.00	33	Mojave Dsrt, CA	2500	0.72	35
Austin, TX	1900	0.95	30	Los Angeles, CA	2100	0.86	34
Baltimore, MD	1600	1.13	39	Newark, NJ	1600	1.13	40
Boston, MA	1500	1.20	42	New York,NY	1600	1.13	40
Chicago, IL	1600	1.13	41	Philadelphia, PA	1750	1.03	39
Cleveland , OH	1600	1.13	41	Phoenix, AZ	2400	0.75	33
Dallas, TX	2000	0.90	32	Pittsburg, PA	1550	1.16	40
Denver, CO	2200	0.82	39	San Francisco, CA	2050	0.88	37
Detroit, MI	1700	1.06	42	Seattle, WA	1300	1.39	48
Fairbanks, AK	1300	1.39	64	St. Louis, MO	1700	1.06	38
Hartford, CT	1600	1.13	41	Sacramento, CA	2100	0.86	38
Honolulu, HI	2100	0.86	21	San Antonio, TX	1700	1.06	29
Houston, TX	1600	1.13	29	San Bernardino, CA	2100	0.86	34
Indianapolis, IN	1650	1.09	39	San Diego, CA	2100	0.86	32
Jacksonville, FL	1850	0.97	30	San Jose, CA	2100	0.86	37
Miami, FL	1850	0.97	25	Tampa, FL	1850	0.97	27
Minneapolis, MN	1650	1.09	44	Washington, DC	1650	1.09	38

To correct for (site latitude - roof slope) angle, first use the table in Fig. 5.15 above, to get the latitude of your solar installation site in USA. Then, to correct for solar climate of the location metro city, as obtained

Fig. 5.16 – NREL Solar-PV Map

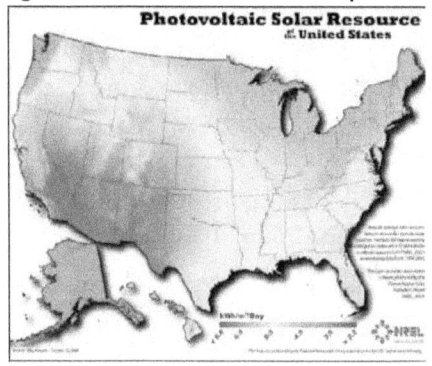

Photovoltaic Solar Resource
of United States

Fig. 5.17 – Correction of: Latitude – Roof Slope

Latit-Slope	Cos(L-S)	Error	Brk Even x
0°	1.00	0%	1
10°	0.98	2%	1.02
20°	0.94	6%	1.06
30°	0.87	13%	1.15
35°	0.82	18%	1.22
40°	0.77	23%	1.30
45°	0.71	29%	1.41
50°	0.65	35%	1.54
55°	0.57	43%	1.75
60°	0.50	50%	2.00

from NREL solar-PV map of the US (Fig. 5.16), use the correction factor obtained from Fig. 5.17. These corrections result in:

BE=Graph x Corr1 x Corr2 (5.1)

where: BE = break even in years
Graph = value from the Fig. 5.14.
Corr1 = Correction from Fig. 5.15
Corr2 = Correction from Fig. 5.17.

Illustrative example: Consider a thin-film, CdTe, solar panel installation costing \$3.00 per watt near Newark, NJ. Assume a flat roof (slope = 0 degrees) on a commercial office building. The latitude for Newark is 40°. Latitude – roof slope is 40°-0° = 40°. Fig. 4.17 indicates the multiplier correction, Corr1, is 1.30.

From the graph, Fig. 4.14, \$0.16 electricity cost Y-axis (NJ) intercepts the Curve C (\$3/Watt) at about 10.5 years.

The second correction, Corr2, is from Fig. 4.15 for the solar climate of the location. For Newark this is 1.13.

So we can calculate the corrected value for break even, BE, from the above equation. Since the flat roof has a slope of 0°, the (latitude – roof slope) is:

BE = 10.5 years x 1.30 x 1.13 = 15 years.

This is not adjusted for any state subsidy or federal tax credit, if available. However, if subsidy is available, BE is reduced in USA by about 30% for a 30% state rebate and/or 30% (more) if a federal tax credit applies. For example, if both apply here, the corrected break even is:

BE = 15.4 years x (1-0.3) x (1-0.3) = 7.5 years. This could have been combined into the original table, but the trend is to discontinue subsidies as the manufacturing and installation prices come down.

Some general trends are observable from the above. For example, it is clearly seen that:

• Electric utility rates below roughly $0.10/kWh "never" permit break even for solar rooftop costs above $3/watt. Roof life expectancy may also be exceeded.

• For states where rates are above ≈ $0.18/kWh, break even is less than 16 years for 2010 costs approximating $5/Watt.

• When rates approximate $1/Watt to $2/Watt (≈ 2015), break even is roughly four years for utility rates above ≈ $0.16/kWh.

• One strategy is to estimate the result using half the electric consumption now at, say, $4-$5/watt, on the southern slope, and the other half in a few years at $2/watt on the eastern and/or western slope. Thin-film, CdTe can be used on the northern slope at a later date.

The tradeoff considerations here involve a straddle between deferring a few years more to install solar rooftop in recognition that the price comes down by Moore's Law vs. not having the beneficial use of solar.

• Re ROI, it is negative until break even, whence it begins to payoff thereafter. Exceptions acknowledged, until solar installations reach $1/watt and only for higher utility rates as shown in the upper left region of Curve A of the graph, would an investor likely go for solar. However, the argument is quite different for non-grid connected installations (Chapter 7) as the motivation and judgment criteria are then very different. Also, because of their MW production, solar farms (Chapter 8) will enjoy shorter break evens and larger ROIs.

If we have made no errors, we have made no decisions

Chapter 6

EMP Protecting the Solar Rooftop

Chapter 6 Overview

This chapter is a continuation of the previous chapter in that an EMP-protected, solar rooftop is now to be added to an EMP protected building. Of course, if an EMP protected building was not done first, as a starting situation, this chapter can not become a stand alone without Chapter 4. For new constructions, both are done with EMP protection at the same time which also provides a considerable cost reduction vs. that if done separately.

Based on an EMP threat, all solar additives must be EMP protected, consistent with the protection addressed in previous chapters. This includes adding shielded solar panels and inverter or micro-inverters, as applicable, shielding all rooftop and other wiring, and the new smart meter to replace the older electrical meter.

Except for protecting solar panels with a low light-transmission loss (less than 5%) of screen solar mesh or other screen shield covers, the preceding chapters also cover many of the EMP protection specifics. They will not be repeated here. Also, remember that this is primarily a lay handbook and detail here is beyond the scope of this handbook. Examples of topics not presented include treatment of roof overhang, guttering, soffits, and dormers.

Because everything ages and because of adverse climate and weather effects, maintenance is addressed. Finally, performance testing of the combined installation includes confirming that (1) the installation shielding effectiveness has been met (see Sect 3.4) and (2) the solar installation meets the performance requirements of the solar portion of the contract.

6.1 Adding Shielded Solar Panels

After placing the framework racks on the shielded roof which holds the solar panels, as explained in Chapter 5, Section 5.4, the solar panels are then added. There are two potential shielding options for providing the 80 dB solar panel shielding:

(1) Shield the panels completely around all six sides with the 20 OPI screen mesh described in Chapter 5. This shielding is to be done back at the factory where the panels are manufactured for quality control, or...

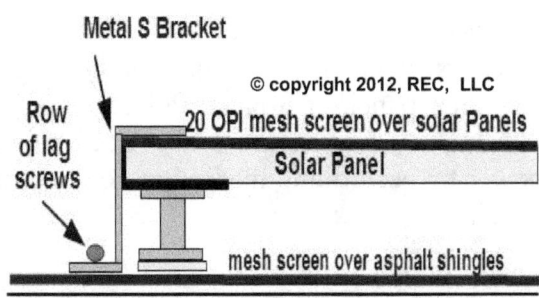

Fig. 6.1 – Wire mesh screen bonding of shielded panels

(2) This method has the advantage of being much less in cost to achieve the shielding as everything gets shielded at once (all panels, panel connectors, micro-inverters, if used, and interconnecting wiring. In this approach, the shield is bonded to the roof shielding over the entire periphery as shown in Fig 6.1. A close examination of the quality of the work is necessary to ensure no leakage points. It is recommended that a second application of the large roll of mesh screen be applied on top of the first immediately thereafter to ensure against any tear and to provide redundant security.

The second approach has the disadvantage of requiring special attention to details and quality of workmanship. All these details are beyond the scope of this handbook.

6.2 Adding Shielded Micro-inverter or Inverter (shield only)

The micro-inverter is a small DC-AC converter which takes a typical 12 VDC output from a solar panel and converts it to 120 VDC or other user load value. As appropriate they are connected in series and/or parallel to produce the desired load voltage. A small micro-converter can be built into the shielded solar panel

Fig. 6.2 – Shielded Micro-converter

near its output connector (this is to be done at the time of solar panel assembly back at the factory) or mounted separately next to the panel as shown in Fig. 6.2 at the right. Unless the entire solar rooftop assembly is to be shielded separately, as discussed elsewhere, the shielded unit in Fig. 6.2 would require a surge suppressor at both input and output cable connectors.

Micro-inverters have several advantages over conventional central inverters. Even small amounts of shading debris or snow lines on any one solar panel, or a panel failure, does not disproportionately reduce the output of an entire array. Each micro-inverter obtains optimum power by performing maximum power point tracking for its connected panel.

Their primary disadvantages are that they have a higher equipment initial cost per peak watt than the equivalent power in a central inverter, and are normally located near the panel, where they may be harder to maintain. These issues are usually surpassed by micro-inverters having higher durability and simplicity of initial installation.

225, 500 kW

10, 13, 15 kW   60, 82, 95 kW   Courtesy of Solectria Renewables

Fig. 6.3 – Shielded Inverter Cabinets

When the total rooftop solar load exceeds about 10 kW, micro-inverters are usually not used and the inverter takes over. The photo at the right shows three inverter sizes ranging from 10 kW to 500 kW. If they were located outside the EMP shielded building they would have to have cabinet shielding of 80 dB plus cable surge suppressors and shielded cabling. To avoid this if possible, they should be placed inside the 80-dB shielded building and existing models used with no further shielding retrofit required.

6.3 Adding EMP Surge Suppressors and Filters

Commercial electronic surge suppressors used for lightning strikes do not clamp fast enough (clamp >100 nsec for lightning; need to clamp less than 5 nanoseconds for EMP, per Fig. 3.4) to protect against the near instantaneous effects of an EMP. Some also may not have great enough current carrying capacity. So it is important that already existent EMP surge suppressors be used for any exterior wire/cable entrance from the outside world to the inside of the EMP-shielded building.

Coaxial Connector
Courtesy of Nex-Tek

Surge Suppressors, wide variety by Protection Group. Photos, not available

3-phase, 200 amp AC Power Line
Courtesy of MeteoLabor

Fig. 6.3A – Two different types of Many hardened surge suppressors

Fig. 6.3A shows two of several EMP surge suppressors available for addressing the wide variety of needs and applications. (The surge suppressors used for high voltage transformers in substations to address the geomagnetic storm needs are all together different and are not shown here)

6.4-Adding the Shielded
Smart Meter

A smart meter (Fig. 6.4) is usually an electrical meter that records consumption of electric energy in intervals of an hour or less and communicates that information at least daily back to the utility for monitoring and billing purposes. Smart meters enable two-way communications between the meter and the central system. Unlike home energy monitors, smart meters can gather data for remote reporting. Such an advanced metering infrastructure (AMI) differs from traditional automatic meter reading (AMR) in that it enables two-way communications with the meter.

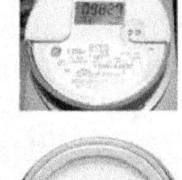

Fig. 6.4: Wire mesh, screen shielded meters

Since the smart meter is mounted on the outside of a building, it will have to be shielded to 80 dB to match the EMP time-domain threat. This can be accomplished by using the same 20 OPI wire mesh engulfing the entire meter on the front and back discussed earlier.

6.5 Testing the Solar Installation (for technical readers)

Contractually, it is necessary in performance tests to demonstrate compliance with the 80-dB shielding requirements over the designated frequency spectrum. This follows the general procedure of Section 4.8 for the building, per se, with some solar additives.

While the building EMP and Solar EMP compliance tests can be done simultaneously, in the early stages it may be best to do each separately. This will facilitate diagnostics-and-fix. The building compliance test must be done first. Details are well beyond the level of this handbook and will not be discussed here.

In addition to the procedure in Section 4.8, a few selected potential inside victims are chosen to be placed in a recording operational mode. Examples include one or more computers, peripherals, cell phones and HVAC temperature monitors for the building and solar developed AC voltage wave forms. *These and maybe others will be monitored only to see if any disturbance is detected.* This will confirm the adequacy of the 80-dB shielding requirement in case any potential victim had not already complied with the applicable 10 V/m radiated susceptibility requirement of Section 3.4. The contractually acceptance EMP compliance simulation is not based on these other potential victims, whose role is to confirm the adequacy of the 80-dB during early stages (first year) of EMP hardening implementation.

EMP-protecting a building is "old hat" to those with clearance who have done this for decades at classified government sites. Since the authors have no security clearance, and nothing appears on the Internet, and we are not knowledgeable of specifics, we are forced to re-engineer expectations so the rest of the civil community can benefit

Chapter 7

Off-the-Grid EMP Protection

Chapter 7 Overview

Seven percent (8 million) of U.S. households (114 million) have a second home. A few percent have an *off-the-path* 2nd or 3rd home, sometimes not connected to the electric grid. Percentage may significantly increase when they learn (1) about EMP and/or other fears of terrorist threats or (2) what can be added to an existing or new structure to make it almost self sufficient. Some of the same features can also be used on a grid-connected installation, but there are differences in content and costs.

Roughly ninety-five percent (95%) of the USA population live on 10% of the land. People cluster because they want to be close to resources, stores, services, jobs, family and friends. However, for the other 10% of the population, the best combination for remote escape may be combining a solar home rooftop installation with an electric vehicle, a propane or gasoline engine-driven electrical generator, and satellite communication devices. They combine this with a highly dependent alarm to provide protection against intrusion, theft and fire. This is not a science fiction, but here today. Moore's Law ensures that prices will continue to come down to entice more participants and budgets.

At the very least, this chapter is a tutorial on what is coming and what it portends in more remote solar installations. At most, some readers may want to do a due diligence and empathize with or participate in change in lifestyle. Product and service providers will seek more info on how to support this sector. Campgrounds, family meeting facilities, state and county parks, fairs and remote resources will increase. Solar's role is dominant. This helps combat unemployment. Remember, if there are 8*% unemployed in 2012 in USA, there are 92% employed, many who will in time be living in increased fear of EMP and other terrorist scenarios, given more press exposure. The 8% estimate may be more nearly 15% when those who gave up looking for work are added back in)

7.1- Need for Solar System Backup

Once the solar site is located away from grid-connectable power, the need for a battery backup becomes paramount. The sun does not shine every day due to cloud cover, rain or snow days. And, of course, during the night there is no sun nor electricity available. Therefore, the solar installation needs to charge up batteries during the day as well as provide the power needed to serve the operating electrical load.

The amount of available chargeable power over a 24-hour period can be estimated from the average equivalent full sunshine hours/day based on the latitude of the installation site. Assume that the panels are facing the zenith (straight up) and no rotatable panel mount (sun tracker) is used or that the panels can be reoriented with the change in seasons. Fig. 7.1, developed by NREL (*National Renewable Energy Laboratory*), a facility of the U.S. Department of Energy for renewable energy and energy efficiency solar map, shows the most favorable locations in US

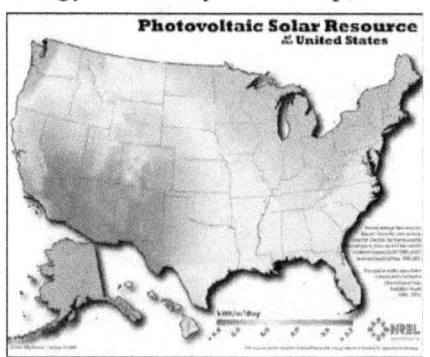

Fig. 7.1 – Solar Intensity map of USA

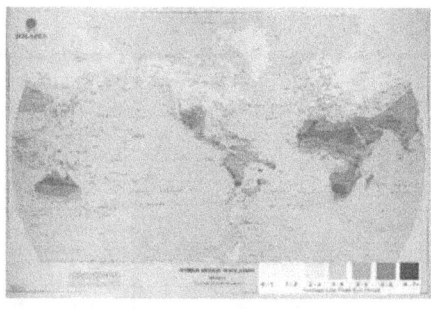

Fig. 7.2 – Solar Intensity map of the world

for solar intensity. Generally, southwest USA has a solar intensity about 40% higher than northern states and some southeastern states. So expect lower solar installation costs in the southwest.

Fig. 7.2 is a companion NREL solar intensity map for the entire globe. Notice most of Africa and Australia are the solar intense areas. Interestingly, Europe has a much less solar intensity even though Europe (Germany, Italy and Spain) has installed most of the solar installations in the world to date. Both USA and China are well behind.

7.2 Representative Off-the-Grid Installation

A representative remote installation is located in the mountains, isolated seashore or any area with low population within a 20 mile (32 km) radius. The median homeowner, second house living space approximates 2,000 sq ft (186 sq m) for a married couple with two children.

Fig. 7.3 shows the general layout of the remote house and its major energy components. On bright sunshiny days, the median solar roof output provides for about 8kW of power. The solar developed energy goes (1) first to service the active household electric load with (2) excess energy being used to charge the electric vehicle's battery banks. The batteries stored energy (appx. 50 kWh) provide the electricity for overcast days and for lower-load nocturnal needs.

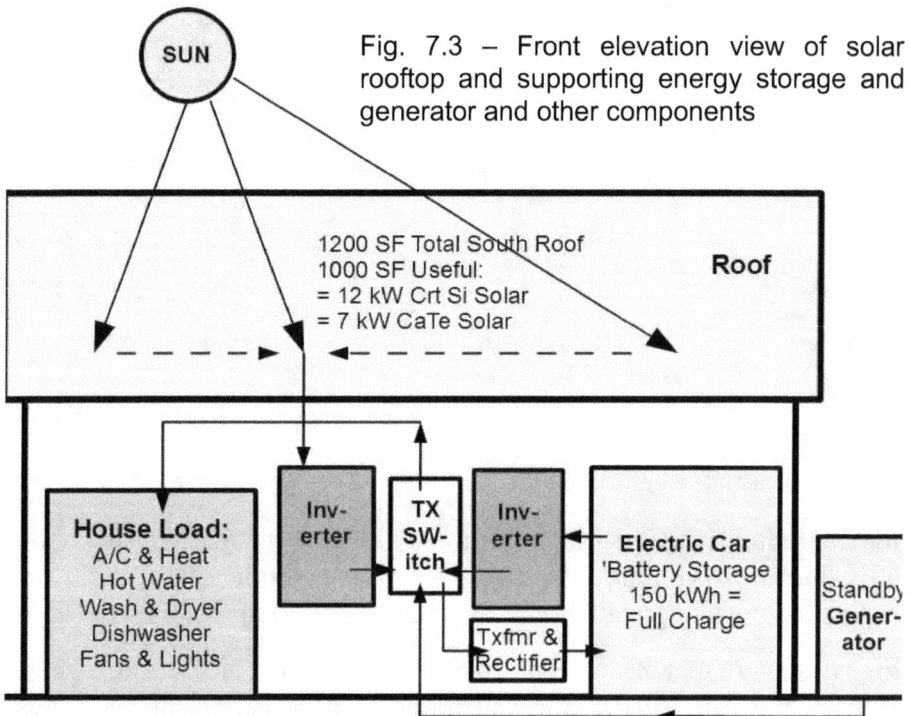

Fig. 7.3 – Front elevation view of solar rooftop and supporting energy storage and generator and other components

Should the electric vehicle (acting as a battery bank) become discharged, the standby generator comes to the rescue. It is typically a 10-15 kW propane or diesel engine-driven generator. Depending on the storage tanks, there may be enough fuel for weeks or months, since the sun will always provide most of the daily needs. It is important, however, to give adequate thought to the size of the fuel tank that may be needed, especially when planning for a worst-case situation.

The rest of Fig. 7.3 shows the inverters which convert the low DC voltage output from the solar panels and the electric vehicles to household 120/240V AC voltage. These run all the appliances and other electric loads. The transformer and rectifiers generate the correct DC voltage for charging the electric vehicle's batteries. The transfer switch connects the selected source for driving the household loads from the three energy sources (1) solar rooftop, (2) electric vehicle, or (3) standby generator.

If enough space in the garage exists, the generator can be located there to avoid the need of an EMP-protected generator as the entire house is EMP protected as described in preceding chapters. In that event the generator is exhausted to the house outside via a waveguide-beyond-cutoff port (defined in Appendix A).

7.3 Calculating Electric Loads & System Needs

Typical electric loads for the remote home are roughly:

Air Conditioning & electric heater = 2500 Watts

Refrigerator = 850 Watts

Hot Water heater = 1,000 W

Dishwasher = 800 Watts

Washer & dryer = 1500 Watts

Five fans = 450 Watts

Fluorescent light bulbs: 300 Watts

Computer & peripherals: 125 Watts

Other (miscellaneous): 1200 Watts

Total: 8,725 Watts

From this list a peak load (remember not all appliances are in use simultaneously) of roughly 7kW exists with a daily average load at the equinoxes (Mar 21 and Sept 21) of about 5 kW. There is still enough extra electricity to slowly charge the electric vehicle.

Latitude, Season Variation and Roof Slope

As stated above, these power numbers mentioned will change significantly with location latitude (degrees North or South from the equator) and the season. For example, the table in Fig. 7.4, fourth column shows that the relevant solar intensity varies over a 1.00/0.35 = 2.9:1 range for the extreme city locations within the U.S. (Point Barrow, AK and Honolulu, HI).

Fig. 7.4– Solar Intensity vs. Latitude

Location	Lat, deg	cos(latit)	Intensity
Honolulu, HI	21.3	0.93	1.00
Brownsville, TX	25.5	0.90	0.97
Miami, FL	25.8	0.90	0.97
Mid USA (KC,KA)	39	0.78	0.84
Madawaska , MA	47.3	0.68	0.73
Seattle, WA	47.7	0.67	0.72
Fairbanks, AK	64.8	0.43	0.46
Pt. Barrow, AK	71.4	0.32	0.35

© copyright 2012, REC, LLC

The solar intensity value is actually based on the combination (cosine of the angular difference) of the slope of the solar installation roof and the latitude as shown in Fig.7.5

Example : If the solar site location were in Seattle WA (48° latitude from Fig. 7.4, second column), then the roof slope should approximate 48° for maximum soar intensity. This is tantamount to an A-frame house construction, with one side facing the south ± 20°.

The location latitude means more solar days of sunshine closer to the equator and far fewer days for high latitudes above roughly 50°.

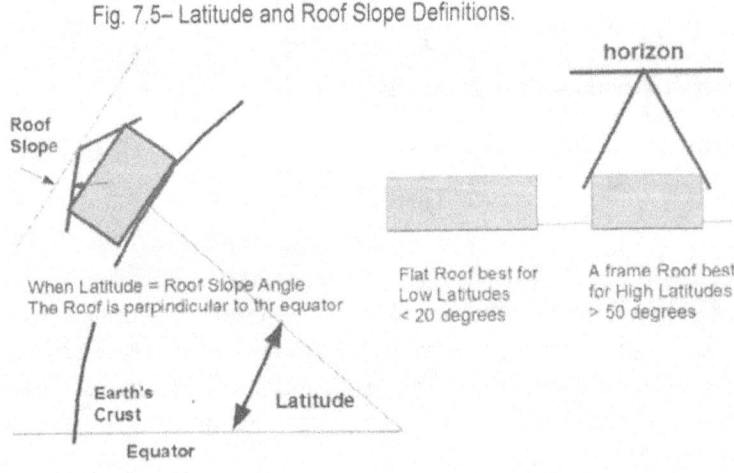

Fig. 7.5– Latitude and Roof Slope Definitions.

There also exist a yearly variation from summer solstice at +23.5° tilt of the earth's axis to –23.5° at winter solstice. These conditions must be added to the angle of table in Fig. 7.4 to get the final variation. Summer angle = +23.5° - Roof slope + Tab. 1 latitude Winter Angle = -23.5° - Roof Slope + Table 1 latitude.

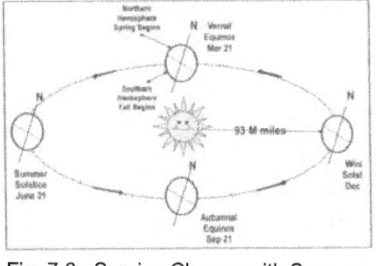

Fig. 7.6– Sunrise Change with Seasons

For fixed solar panels (do not rotate or move), best performance exists when latitude location = roof slope angle. However, because of longer solar summer days, the optimum favors summer angle pointing.

7.4 Battery Backup or Electric Automobile

Batteries are essential for solar substitute or backup, either in the form of lead-acid storage or lithium-ion. (Lithium-ion batteries are about five times the cost per kWh (kilowatt hour) as lead-acid types). Batteries come into use at night when the sun services elsewhere and during overcast days. Should the home owner also have an electric vehicle, this may substitute for the battery bank, per se, provided arrangements have been made to add a special access and transfer

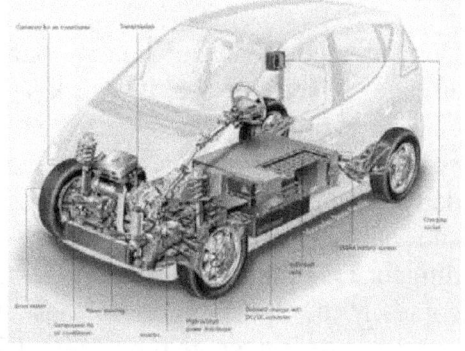

Fig. 7.7– Typical Electric Car, 2012 model

switch between the solar source and the automobile (Since this use will likely violate the car battery guarantee coming with   the automobile, it may not be a viable option. Perhaps, this may change in the future for some consideration).

Since the vehicle may be frequently used for transportation, daily connections and disconnects can become both a nuisance and a bit dangerous. Note: don't worry about 12V and 24V DC (unless your hands are wet) as you will not feel the potential shock. That's why the telephone company uses 48 volts DC on their customer distribution lines for over 100 years since it is near the threshold of trickle perception. But, remember this: The automobile must also be EMP protected or it too will become dysfunctional from an EMP incident if it is not inside a shielded facility as well. Therefore, electric vehicles may be ineligible.

As explained below in generator backup, the average home size consumes roughly 30 kWh (kilowatt hours) of electric energy per day. Assume the storage capacity is only required for one day (the standby generator takes over for longer periods and to recharge the batteries). Then, the batteries must be capable of supplying 30 kWh before a recharge.

A typical 12-volt, automobile lead-acid battery stores about 1,000 watts (based on 85 Amp Hours). Unless recharged during the day, it would take about 30kw/1kw or 30 such batteries to substitute for solar in an average day. Adding the price of purchasing and installing an inverter, transfer switch and all the wiring would approximate a total battery backup cost of $3,000 to $5,000 cost.

Compare the lead acid battery bank investment of the above with a storage capacity of roughly 50 kWh for an average lithium-ion battery electric car where the battery cost about $600/kWh. The electric car covers the daily storage needs above. But, it also requires the other daily wiring connections and disconnections requirements mentioned and battery guarantee. So, which is better. Obviously, further due diligence is required and the inclinations, health, etc of the homeowner come in to play. So, no further discussion here. Use the lead-acid battery as default since it cost about 15%-20% of the lithium-ion battery cost.

7.5 Generator Backup

As shown in Fig. 7.3, the combined solar-battery installation may still need additional electricity backup, especially for multiple days in a row of overcast skies. Here the electric, backup generator comes to the rescue in one of five different forms of energy source: propane, natural gas, LPG (liquified propane gas), or diesel.

Fig. 7.8– Typical 15 kW backup generator

USA generator suppliers include Coleman, Generac, Kohler, Guardian, Onan, Briggs & Stratton, GE, John Deere and Caterpillar. See Fig. 7.8.

Engine generator sizes range from 10 kW (low end for homes and for less than full load) for residential use to over 1 MW for larger commercial consumption. Consumer needs are approximately 0.2 x normal daily kWh (kilowatt hours) consumption, based on five hours full daylight climate. Fig. 7.1 is the NREL solar map of USA. Notice that in southwest USA (CA, AZ, NV, NM, TX and CO), the required generator size may he reduced by approximately 25% of the above. However, for the New England states and NY, PA, OH, MI, WI and northwest OR and WA, the generator capacity may need to be increased up to 25%.

Prices for engine generator sets approximate US$0.20/watt. So the popular 10 kW to 20 kW home standby generators would cost approximately $2,000 to $4,000, respectively. Delivery may be free, but installation is extra including wiring up the transfer switch and connecting to the fuel storage tank or access supply. Almost any size generator can be rented for emergency needs or long-term use. Rental companies are readily found on the Internet.

The supporting generator fuel supply needed depends on the type of fuel mentioned above, accessibility and the degree to which an after an EMP event will linger in availability from regional suppliers. This may range from a few days to months. Discussion here, often a matter of mental concerns, is beyond the scope of this book.

Generator only backup

Since some generators can have a near instant startup time, an argument exists for eliminating the batteries altogether. The savings can be put into additional fuel storage tanks plus fuel. However, an argument exists for the increased insurance offered by a partially redundant backup system.

Large Generators for Big Buildings

Previous discussions address homes and small buildings requiring generator backup power, usually under 100 kW, but up to 500 kW. What about larger power for larger buildings?

Fig. 7.8A is one example, illustrating a 4 MW source (enough power for 700 average homes). This Caterpillar Cat® 20-cylinder C175 features the greatest output of any high-speed diesel generator set, which makes it practical for mission- critical standby power installations. This 20-cylinder model generates more power from a smaller footprint, requires

Fig. 7.8A. A Cat® C175, 4 MW, 20-cylinder Diesel Generator by Caterpillar

less maintenance, and is certified to meet EPA Stationary Emergency Tier 2 emissions limits. The C175-20 is available with factory designed and tested attachments that offer flexible packaging options for easy installation and quality performance.

7.6-Off-the-Grid Cost Expectations

To get some 2012 sense of what it might cost to invest in an off-the-grid, EMP hardened home. The estimates are made for the years 2013 and 2020 as shown in See Fig. 7.9. They are made for both an existing home and a new one of the same specifications, yet to be built. Estimates are presented for the EMP hardening as discussed in previous chapters, plus the solar rooftop installation for 8 kW power, which should be more than adequate for a 2,000 sq. ft. home. To this is added the cost of a battery bank, propane generator and buried tanks.

Fig. 7.9 shows that the cost to EMP protect the 2,000 sq ft. home including protecting a solar installation It is roughly 21 percent of the home value in 2013 and falls to 20 percent in 2020, mostly because of the projected savings due to mass production in the next seven years. These costs are less for new homes, yet to be built, because of the easy access to bonding all corners of a shielded building and related access.

Fig. 7.9 – Costs estimates for EMP survival, off-the-grid, 24/7 electrical/electronic operational home

Cost Estimates for Remote, Off-the-Grid, EMP hardened Facility

Year		2013		2020	
Item Identification	Notes	Existing	To Build	Existing	To Build
Square feet under air		2000	2000	2000	2000
Square meters		186	186	186	186
Market Value, $US1k	A	250	250	250	250
EMP Harden, House	B	71	48	50	33
8 kW Solar	C	20	20	8	8
Battery Bank	E	3	3	3	3
Propane Generator	F	5	5	5	5
Propane storage tanks	G	2	1	2	1
Total Cost, $US1k		351	327	318	300
Cost for Solar in $US1k		20	20	8	8
Cost for Solar in % add		8	8	3	3
Cost for HEMP in &US1k		71	48	50	33
Cost for HEMP in % add		28	19	20	13
Cost for Remote add in %	H	40	31	27	20
Item Identification	Notes	Existing	To Build	Existing	To Build
Year		2013		2020	

Notes: Assume: No Inflation from 2013 and 2020
  A: Assume Market value of house and property = $125/sq. ft
  C: Solar cost installed = $2.00/watt in 2013
  H: Total remote extra = Solar + HEMP Harden + Battery + Propane + Tanks

It is important to understand that the above estimates are subject to significant variations since, for the most part, this has yet to be done. Its do-ability is not questioned since essentially all of the shielding, grounding, surge suppressors and filters already exist or are adaptable within the engineering community.

7.7 Other Survival Provisions

It is also important to understand that the above discussion permits a 24/7 availability of (1) the electronic and electrical devices within an EMP-protected home or building, plus (2) the EMP-protected solar rooftop, batteries and generator. Both are needed as the assumption is that the electric grid has burnt out with an EMP incident and it may be months or years before restoration. But what about other provisions?

The telephone, radio, TV, cell phone, and other such telecommunication devices have remotely located transmitters which may be damaged or burned out as well in an EMP incident. Therefore, the homeowner cannot use his matching transceiving devices. Here is where an emergency short-wave (EMP protected) radio is vital as the homeowner will/may find out from surviving local and long-distance broadcasts stations what emergency services are available (fire, police, medical/hospital). He will also learn about what available retail facilities are operational for supplying food and water, generator fuel replacement, and the scores of other everyday items of daily living. Remember, an EMP incident does not physically destroy infrastructure and kill humans (see Chapter 1). But, it can make people become desperate.

The refrigerator, with freezer is a big food and drink storage unit, operational 24/7. Remember the big electrical loads are hot water heaters, air conditioning, washer/dryer and electrical cooking. So the reader should review Appendix B to practice electricity conservation, especially if something goes wrong. Example: Suppose the backup generator was low on fuel supply (say three days) and there turned out to be an extended low pressure forecast in the area with a week of overcast skies. No solar output. Then what?

Fig. 7.10 – Organization diagram for priority planning using optimization criteria and six building/sectors to be protected for human survival.

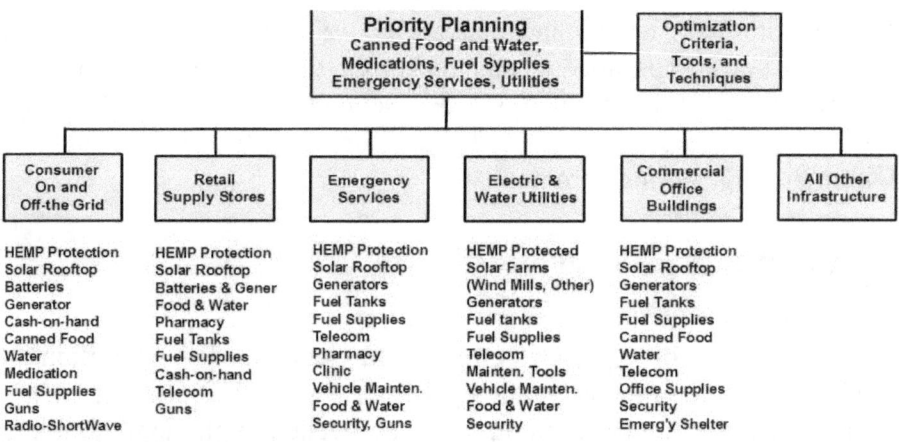

Fig. 7.10 is one organization diagram used to illustrate some of the many priority plannings, optimization criteria, and facility structures and services that impact survival. Priority starts at the left and moves to the right. Human beings living in their home come first. The many retail supply stores comes next since the basic food, water, medications, and fuel provisions determine how long humans can survive beyond their own stored supplies. The emergency services include the fire, police, medical and related support.

There are dozens of available remote survival books from Amazon, other book suppliers, bookstores and libraries that address survival provisioning. Many are located on the Internet as well. So there is no reason to cover that topic in this book.

7.8 The EMP Role of Transportation

The substantially missing topic in the previous section and in Fig. 9A is transportation. Without this, nothing will arrive at the distribution centers and the retail outlets for consumers to replenish their waning provisions, which also depends on vehicles.

Four tiers of transportation are needed:

1. Home vehicles to pick up the replacements at retail outlets
2. Trucks to delivery retail goods from distribution centers to retail outlets
3. Trucks and railroads to deliver goods to distribution centers
4. Aircraft and ships from regional centers and foreign countries.

Figure 7.11 table shows some of the major areas where EMP protection is needed. Perhaps, the most critical are the item-delivery vehicles, Column B. Without re-plenishment from vehi-cles, the shelves of food, drink, and pharmacy stores and fuel stations will remain empty and millions will starve to death or die of needed medication.

Fig. 7.11 - General Priority Identification for Civil Commercial, Industrial & Residential EMP Hardening Applications							
		A	B	C	D	E	F
Services		Home	Delivery	Stores	Supply	Railway	Internat
Category	Specific	Vehicles	Vehicles	or Bldg	Centers	Engines	Airports
Survival	Food & Drink						
	Pharmacies						
	Gasoline/diesel						
Utilities	Electric						
	Water						
	Gas & Other						
Emergency	Fire						
	Police						
	Personal						
Other Priority Buildings	Telecommunicat						
	Shopping Malls						
	Comm Office						
	Motels/Hotels						
	Selected Homes						
	Wealthy Homes						

Note: See Figs. 7.9A and 9.1 + discussion

Home vehicles, Column A, are needed by con-sumers to pick up replenishments from retail outlets. It may be prudent to have an extra electric golf cart, as backup.

Vehicles are also needed to service the distribution centers from the regional area depots and from airports and shipping ports. All vehicles in the survival chain from home use to supply centers will have to be retrofit with kits to protect the engine control microprocessors and selected wiring shields with surge suppressors. This also applies to the Transmission Control Unit, the Anti-lock Braking System and body control modules In new manufactured vehicles in quantity, this may be attained for perhaps less than $500/vehicle.

Parenthetically, it is also noteworthy that in William Forstchen's, "One Second After", *New York Times* bestseller book on life after an EMP incident, that the vast majority of premature deaths resulted from food starvation, water shortages and no available medication and fuel, because supply vehicles were rendered inoperable. Exceptions were the few pre-1982 trucks in service where micro-processors were yet to be used.

One closing comment has to do with surviving, in general, and remote locations, in particular when refrigerated food supplies, and/or water and gasoline run low. If your EMP protected home is the only home in the area with refrigerator, lights and fan, other desperadoes may chose to attack you for the basics you still have. It may help to cluster nearby with a few other homes and folks of your ilk for sharing expenses, self preservation and protection.

One of the authors (Donald White) went through Hurricane Charley at Lake Suzy, Florida on Friday, 13 August, 2004. We were without electrical power for 13 days. No cell phone service for five days as the local cell tower was down without electricity. The nearby Publix food chain ran out of fresh food and its refrigerators had no generator backup. Gasoline pumps at local gas stations had no power and so on…. When supplies run out, our fellow man can become very ugly. They may fight each other for who gets at the store shelves first.

As the post hurricane period developed, we were the only folks in the area with a generator and an extra refrigerator in the garage. The neighbors "loved us". Many slept in our home as we had fans going (not enough power to operate the A/C) 24/7. Fortunately a drive to North Port, eight miles to the north, was spared by Hurricane Charley. We made this run daily.

Instead of a hurricane, suppose this had been an EMP incident, instead, in which all electricity and electronics were wiped out for 500 miles. What then? Fig 7.12 shows the distance on the ground (line of sight) impacted by an EMP incident vs. altitude for a non-mountainous location (sea level).

Fig. 7.12 Ground Radius EMP Destruction vs. EMP Explosion Altitude

Altitude		Ground Radius		E Field*
Miles	Km	Miles	Km	kV/m
30	48	563	900	
38	60	633	1,014	40
63	100	816	1,306	48
125	200	1,149	1,838	51
188	300	1,409	2,254	57
300	480	1,780	2,848	

* See Fig 1.3 re Field Strength

The low end of EMP altitude (about 30 miles) suggests that ground electric burnout distance is 500miles/8miles = 63 times further away than the North Port example above in the Hurricane Charley incident. In other words, unless one had stockpiled provisions, there was little hope for survival unless many other supply stores were similarly EMP protected. Perhaps, that could begin with EMP hardening modified warehouses, such as *Costcos* and *Sams Clubs*; then selected *Walmarts* and *Targets*, etc
.
But, who is to pay for this and how? See Chapter 9.

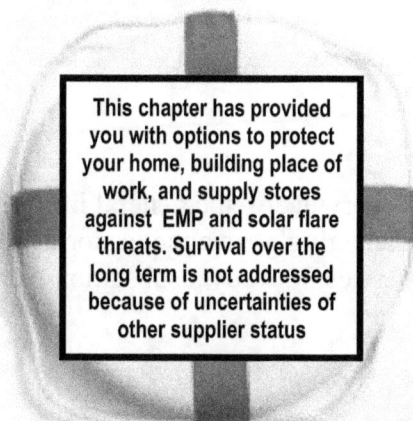

This chapter has provided you with options to protect your home, building place of work, and supply stores against EMP and solar flare threats. Survival over the long term is not addressed because of uncertainties of other supplier status

7.9-What if this had been a Solar Storm instead of a EMP?

The results would be very different. While both EMP and a strong solar flare that produces a severe geomagnetic storm may take out the electric grid in a designated area, a solar storm would not immediately stop transportation; and all electronics and electrical devices would still work if they had backup power. If the grocery stores, pharmacies, gasoline stations and water and sewer utilities have back-up solar power (even if unshielded in the case of a solar storm) then life is not severely impeded within the community, especially if other many communities across the country also have solar backup power.

Fig. 7.13 *General identification of EMP and Solar Flare threats and survivors*

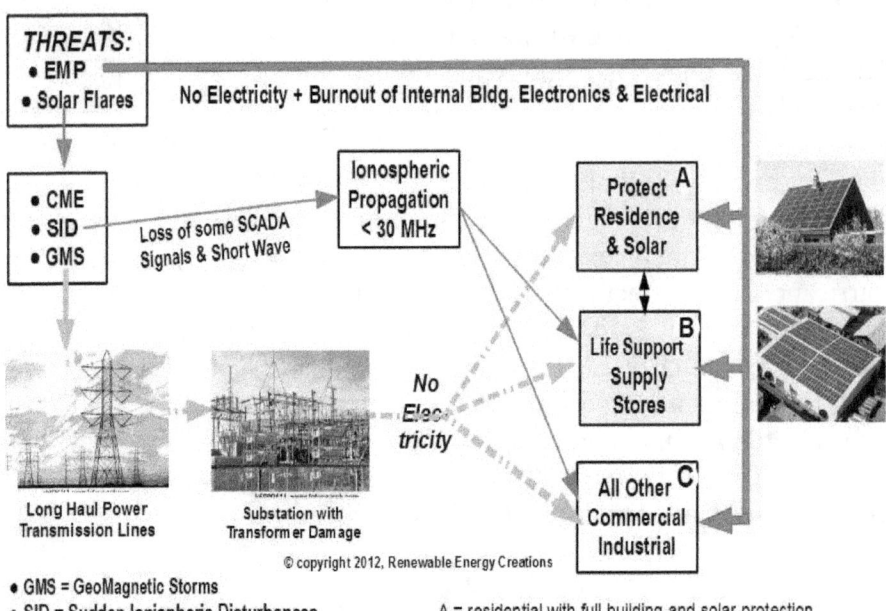

Fig. 7.13 presents an overview of the general impact of an EMP incident vs. a lesser damaging geomagnetic storm; although, unless protected, the solar flare can destroy a substation transformer servicing a long transmission line.

Epilogue to Chapter 7 and Notes on Vehicles

As mentioned earlier, chapter 7 is a somewhat stand alone chapter since it builds upon and exemplifies the applications of previous chapters. In those previous chapters, discussions on products and techniques to harden against EMP bursts, while providing for ongoing electricity independent of the failed electric grid, were described.

Section 7.7 touched upon the survival importance and priority of EMP-protected vehicles (or immune pre-1982 vehicles), barely giving one-half page to the discussions, since the subject is beyond the topic of EMP *building* protection of this handbook. A planned future volume in this series will address the problem of EMP protection of vehicles.

Meanwhile, it can be said that the main protection required of most vehicles is to protect the microprocessor based engine controls by additional shielding the content box and placing surge suppressors at all associated wires (signal leads and DC power) on the housing connectors. A sturdy grounding strap is made between the housing shield and the metal frame of the vehicle. This is repeated, as necessary, with several other circuits.

It may also be necessary to replace one or more cables leading from or to the microprocessor shield with shielded cables. EMP simulator chamber testing should be used on the retrofit vehicle to determine the adequacy of the EMP hardening and the margin of protection. This will be helpful in finding the most economical tradeoff to avoid over or under EMP protection.

Extensive simulated EMP testing on vehicles has been done at White Sands Proving Ground. The Nuclear Effects Directorate has tested many automobiles and trucks (Ref: EMP Effects on Vehicles *by Jerry Emanuelson, Futurescience, LLC)*.

The United States EMP Commission also preformed EMP testing on vehicles. Based on that small sample of pre-2003 vehicles, at higher field levels, 70 percent or more of the trucks on the road will manifest some anomalous response following EMP exposure. Approximately 15 percent or more of the trucks will experience engine stall, sometimes with permanent damage that the driver cannot correct. Similar to the case for automobiles, the EMP impact on trucks could trigger vehicle crashes on U.S. highways. As a result, many more vehicles could be damaged than those damaged directly by EMP exposure.

Here are some additional recommendations for vehicles from the web site www.emp-safeguard.com:

For vehicle owners:

• Until the vehicle EMP susceptibility data base becomes far greater, the following will not eliminate EMP susceptibility; but will help reduce it.

• Park your vehicle in a sheet metal shed or steel building to further mitigate susceptibility. Further improvement would result from sheet metal placed over the floor. (acts as a HF/VHF capacitor to short the induced EMI/EMP)

• Temporary storage of vehicle: Ten mil sheet metal floor. Cover vehicle with some 18" - 24" overlapped. 1-3 mils (household aluminum foil). Secure overlap with duct tape and connect foil to mat flooring with widths of foil bonded to each with duct tape. (This is a crude temporary shield).

• In the northern latitudes above roughly 40 degrees, engine oil warmers are plugged into AC outlets to make starting the vehicle easier in cold winter in the following morning. Use cable surge suppressors and ferrite absorbers at the cable auto entrance points.

• Do not count on any significant susceptibility reduction if vehicle is parked in your resident garage, since walls are radiation unprotected and all wiring enters at the breaker panel from an octopus-like antenna house circuit wiring EMP radiation pickup system in the outer house walls.

• The greater the vehicular sheet or foil or deposited metal vs. plastic skin overage of engine and cab, the lesser is susceptibility. Are microprocessors metal foil or metalized covered and protected with surge suppressors? Is the cabling shielded and clamped to the metal frame?

• The problem with the above suggestions is that that there appears to exist no sound quantitative documentation on vehicle EMP susceptibility. This is especially unfortunate as the United States is a litigious society. Perhaps the AMA might nudge its members to do this by (1) establishing test set-up conditions and test procedure standards, and sharing costs and results. Meanwhile, some of the above EMP vehicle recommendations may represent overkill; and some are still inadequate for EMP protection. This is an unthinkable and unprofessional situation.

For vehicle manufacturers:

• Design and test selected vehicles to meet MIL-STD-188-125 field strength 50,000 volts/meter EMP limits.

• Provide a retrofit kit and services to reduce vehicle radiated susceptibility. Some measure of the safety margin must be reported along with its meaning by the manufacturer.

• Provide vehicle owners with an EMP manual of vehicle use and parking recommendations.

• The greater the vehicular sheet metal vs. plastic skin overage of engine and cab the lesser is susceptibility. Cover or coat microprocessors with metal foil or metal paint or deposition.

• Protect cable entry and exit with surge suppressors. Selected cables should be shielded. Cables should have ferrite clamps at both ends. Keep cables close to metal vehicle frame.

• Since air gaps in metal parts break down at 3000 volts per millimeter potential difference, insulation of some wiring and especially motors, generators, starter relays, and ignition coils are vulnerable. Increased EMI protection is needed.

• **The 80-year-old practice of grounding the negative terminal of the battery to the vehicle frame and using the frame as the return in circuits should be ended.** The reason is that the circuit loop area is somewhere between 100 to 10,000 times (60-100 dB) greater than a simple twisted wire pair. This means that all the potential electromagnetic interference picked up from coupled transmitter radiations, local and distant, from radar and especially from EMP will be reduced accordingly. For less than $100 in additional wiring plus labor, the EMP hardening impact will be many times rewarded. (Remember you heard this here first).

Chapter 8

EMP Solar Farm Hardening

Chapter 8 Overview

This handbook has addressed solar rooftop installations such as applied to commercial, industrial, and residential installations all supplied by the electric grid. Even remote sites, where no grid exists, are presented next. However, the vast majority of total solar installations, measured by MW or GW of solar power generated become the province of the electric utilities and some private investor group owners selling back electric power. These installations are called *Solar Farms*.

Solar Farms are what this chapter is about. They may range in size from roughly 10 acres (producing 1 MW) to the GW range covering many thousands of acres.

Solar-PV farms operate in a similar manner as solar rooftops, needing solar panels, inverters, interconnecting wiring, etc. The big difference in solar farms is that the solar panels are usually mounted on solar trackers, equatorial mounts, which rotate in azimuth to track the sun over at least 90° of its diurnal path. All the above components must be shielded, with most of the cabling being buried in conduit for additional EMP hardening and periodic surge suppressors as a vital component.

The command and control of the entire solar installation and the interface with the electric grid initiates from the central control building(s). This facility requires EMP hardening as previously described for EMP hardening buildings.

8.1 Solar Farms

A solar farm includes a solar-PV panel installation, usually mounted on the earth's surface, which starts in size from roughly 1 MW (capable of serving about 125 residents). However, their lower end of size is more likely greater than 10 MW (serving roughly 1,250 residents). Some are government-owned, such as the one at Nevada's Nellis Air Force Base, while most are owned by electric power companies or private investors.

A few solar farms are shown in Fig. 8.1

Fig. 8.1 – Four representative photovoltaic solar farms.

In actuality, there are three types of solar farms:

(1)-*Photovoltaic* in which solar-PV panels are usually similar to those installed on rooftops to capture sunlight. These are the main type of solar farms discussed in this book.

(2)-*Parabolic troughs* are large arrays of parabolic mirrors or reflectors which collect heat energy from the sun and refocus the radiation onto a receiver (absorber) tube located at the focal point of the parabola array (Fig. 8.2). A heat transfer fluid is heated to high temperature as it circulates through the receiver tubes. The heated fluid is then piped through a series of heat exchangers where it releases its stored heat to generate high pressure steam. The steam is then fed to a traditional steam turbine generator where electricity is produced.

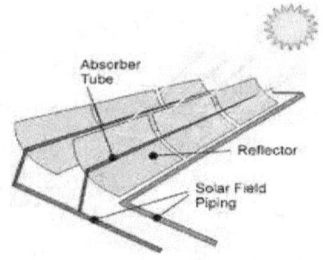

Fig. 8.2: Parabolic Trough

(3)-*Power Tower* consists of hundreds or thousands of tracking mirrors which reflect sunlight to an oil heater on top of a tower which boils water to generate electricity.

Parabolic troughs and power tower type solar farms are shown in Fig. 8.3. These types are discussed further in Vol. 2 of this EMP series to debut in late 2013. They will not be discussed further in this book.

Fig. 8.3 – Four representative, non-photovoltaic solar farms.

Cost reduction is a major limiting factor in the development of solar farms. Large-scale operations cost in the hundreds of millions of dollars, but prices are expected to drop continuously as quantity of production increases and technology continues to improve according to Moore's Law.

Land use is another concern. As previously shown in the NREL solar map, the U.S. Southwest receives strong year-round sunlight, making it ideal for solar farms, but much of the available land is government-owned with overlapping jurisdictions. The land is usually isolated, which requires transmission lines to run long distances to deliver the power to where it's needed.

Since many are concerned that solar energy does not work at night and on cloudy days, most facilities can store energy in reserve for these periods. Solar farms can also be used to provide voltage control to wind turbines, allowing more wind energy to enter the grid at night, when solar farms are dormant.

A Few Specific Examples

Solar Millennium, LLC and Chevron Energy Solutions, the joint developers of this project, propose to construct, own, and operate the Blythe Solar Power Project. The project is a concentrated solar thermal electric generating facility with four adjacent, independent, and identical solar plants of 250 MW nominal capacity each for a total capacity of 1 GW nominal. It is important to note that 1 GW is the entire solar power produced in USA in 2011.

The applicants have applied for a right-of-way (ROW) grant from the U.S. Bureau of Land Management for about 9,400 acres of flat desert terrain. The total area that will be disturbed by project construction and operation will be about 7,030 acres. The area inside the project's security fence, within which all project facilities will be located, will occupy approximately 5,950 acres.

Skyline Solar is another, but of different character. Like a few others, Skyline plans to reduce costs by concentrating sunlight onto smaller solar cells. Skyline Solar, however, says that it can better compete with other energy sources by combining two technologies that can be produced in high volume using existing equipment and that have been demonstrated in the field for decades: conventional silicon solar cells and reflective parabolic troughs, which are used now in solar thermal plants. Skyline Solar has replaced those tubes with narrow solar panels, adding a heat sink to keep them from getting too hot. The troughs concentrate the light by about a factor of 10, increasing the power output of the panels by about the same amount as conventional solar panels without concentrators.

EMP protection of these non-solar-PV farms is not addressed in this handbook. In general, it is likely that EMP hardening should be less expensive as there are far less solar sensitive devices or areas to address in this type of solar farm per MW of installation.

8.2 Shielding and Grounding Solar Trackers

Solar farm sun trackers are azimuth (east to west) rotating mounts holding solar-PV panels or parabolic troughs. They track the sun from roughly 45° in ascending solar elevation to 45° in descending solar elevation for a 90° solar pointing coverage for maximum output. Thus, averaged over the year, six hours of maximum daylight power development are possible unless the climate has frequent overcast skies.

There is a temptation to push the 90° tracking arc to greater angles of coverage for more output but there are two limiting factors:

(1)-More real estate is required because adjacent solar panels will block or eclipse its neighbor at lower elevation angles so that the power density output is actually reduced.

(2)-Lower elevation angles look through greater atmosphere and atmospheric scattering and absorption make the solar panel pick up less efficiently for converting sun light into electricity. However, it may be financially viable to consider up to 120° coverage in the optimization process, especially if the land acreage is inexpensively available.

Fig. 8.4 – Multiple Solar Panel Mounts

EMP Protection

How do we protect the solar panels and their azimuthal tracking mounts from an EMP incident? Previous chapters discussed in general the wire

124

mesh screen shielding of solar panels. Those techniques can be adapted here in a modified form as well.

An option may exist for using unshielded solar panels, but spreading the wire mesh shield roll over a group of panels on each rotating mount. This "slip-on sock" shield would have to cover all six sides of the rectangular solar panel sub-array and mount configuration and include its micro-converter. So the output of each mount, having from perhaps four to 50 solar panels (see Fig. 8.4) would be 440 VAC (higher voltage and lower current for large solar farms) and routed immediately via shielded cable to a buried shielded cable conduit.

The shielded mesh-sock concept must be secured at the sixth side where the mesh coverage meets. Either a conductive gasket, binder-clip devices, or shielded zipper-tubing is used to secure the mesh-on-mesh interface to protect the integrity of the 80-dB shielding. It is not quite this simple, though, since the mesh must interface using gaskets with the mounting hardware.

The motor drive on the end of each solar mount tracker would also be similarly shielded including its power source cable. It can be contained in the same solar panel shield.

8.3 Interconnecting System Wiring

Control signals exist to each mount group from the central solar farm control center. Naturally all these communication cables are enclosed separately, but in the same shielded buried conduit. Surge suppressors should be periodically used. The intricate details are well beyond the scope of a general handbook like this one.

In very large solar farms, several trackers are grouped into sections where the buried conduits meet in a shielded well; and the voltage is stepped up for greater efficiency for transporting power over longer distances, say roughly 0.5 km. Whatever the design, all are shielded to meet the 80 dB criteria discussed in earlier chapters.

The point of the above discussion is that the technical considerations lend themselves to retrofit in updating existing installations, thanks mainly to the 20 OPI wire mesh covering the solar panels and their micro-inverters, as discussed in earlier chapters. However, installing EMP hardening is far less expensive when carried out at the time of a new solar farm installation.

8.4 EMP Hardening the Central Control Buildings

For shielding the solar farm central control building(s) follow the same procedure as discussed in detail earlier. All of the basic engineering principles are the same, except that the buildings will be different in size and content. However, remember that the building has a companion utility switching yard connecting the output to the electric grid. It is less complex than a substation as the input/output voltage ratios are much less. Details will appear in Vol. 2.

8.5 Solar Farm EMP Performance Testing *(for technical readers)*

An earlier chapter provided some insight as to how EMP performance testing is done to corroborate the 80-dB shielding effectiveness compliance. MIL-STD-188-125, Fig. 1.1 discussed in Section 1.4 provides the setup shown in Fig.8.6. For a solar farm, however, the transmitter needs to be set up in an overhead location, preferably in a blimp. A helicopter will do, but is less favorable. Once larger hover-drones are developed and approved for civilian use, these may become a suitable platform for the remote-controlled transmitter.

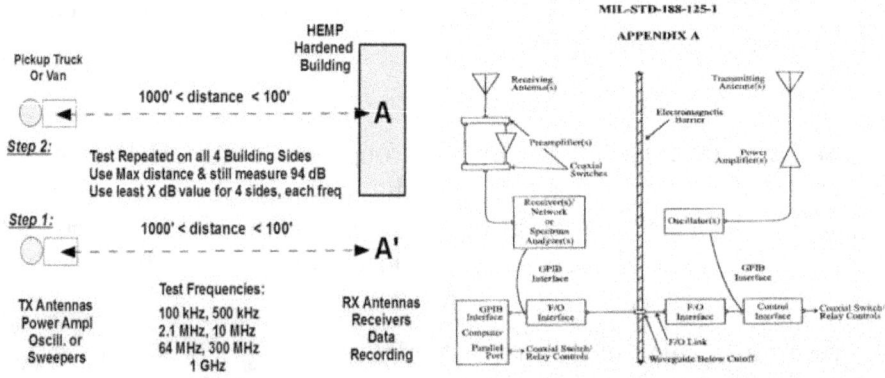

Fig. 8.5 – Test setup for determining shielding performance

Fig. 8.6 – MIL-STD-188-125, Fig. A.1 Shielding Effectiveness Instrumentation

The two illustrations shown in Fig. 8.5 and 8.6 are similar in concept except that MIL-STD-188-125 reverses the role of the transmitter (TX) and receiver (RX) which is applicable except for large enclosures. Fig, 8.6 is more automated than Fig. 8.5 and is a the preferred method. However, the law of reciprocity may be violated because for technical reasons.

8.6 – Epilogue: Tracker System Design Efficiency

Although not directly connected to the topic of hardening Solar-PV farms from an EMP incident, the matter of tracker system design efficiency needs to be addressed because of some questionable practices. For example, some farms are designed for the trackers to rotate between 45 degrees on either side of the zenith (sun at high noon). Why not go further and save money. Fig. 8.7 shows some data.

Fig. 8.7 - Fixed vs. Tracker Performance, Efficiencies & Yield for Solar-PV Farms for Stipulated Assumptions

Fixed or Rotat	Fixed	Rotational	Rotational
Rotat Angle	Horizontal	45°-90°-45°	30°-90°-30
Rel Efficiency	63.7%	73.6%	82.9%
Improvement %	Ref = 0%	15.6%	30.2%
PV Farm Size	100 MW	58 MW	65 MW
Annual Output	$27.4 M	31.6 M	35.6 M
Rel. $ Savings	0	$4.2 M	$8.2 M

Assumptions: 1 kW-hr = $0.15/hr
1 kW produces 1,825 kW-hr/annum
Relative efficiency vs pointing at sun from sunrise to sunset

A 100 MW solar farm is selected as the example. While there is an extra $4.2M/annum gross revenue production/annum in using a traditional 45°-90°-45° sun tracking system, why not go for 30°-90°-30° tracking setup and gain an extra $4M/annul ($8.2M - $4.2M) in revenue? Some may say it takes more acreage. True, but is another 100 acres going to cost $4M/annum? Of course not. This and several other related practices of some solar farm designs should be questioned. These matters need to be addressed first (or independently) in addition to the concept of EMP hardening of solar-PV farms.

Chapter 9

EMP Protection & Backup Power Cost Estimates

Chapter 9 Overview

Cost estimates were presented in chapter 7, but only for solar-PV rooftop installations and for battery and generator backup support for small buildings and homes. Little has been presented on costs for EMP hardening of the same structures. This chapter provides some early cost estimates.

Because this book is written about unclassified EMP hardening of the civil world of building infrastructures and homes, there is yet no history to use as a data base. However, we can present some estimates on same based on familiarity with the necessary components. Also, costs in 2013 will be much higher than 2015-2020 costs when the benefit of quantity production has set in. So, both time periods will have to be addressed.

Finally, the matter of retrofit of existing buildings and homes vs. new infrastructures is addressed. Costs for adding EMP hardening for existing structures may range from 1.4 to 2 times those for new structures. The reasons ares (1) difficulty of building surface access, especially for the building foundations, and (2) the esthetics of installation appearance.

9.1- Structure Surface Area, General Considerations

Cost estimates begin with determining the total outside surface area of the structure to be covered with shielding materials, This is usually divided into two parts based on living and storage areas. Total area includes foundation, roof, and four (or more) sides. This area is divided into (1) outside windows and doors, and (2) total external area less windows and doors.

The total surface area, depending upon the building, may include elevator and HVAC roof shacks, house dormers, basements and garages. If any are not considered, remember that any electrical or electronic devices in those spaces will be unprotected. Otherwise, they will be burned out or damaged in a EMP incident. This also applies to the microprocessors used in vehicles parked in the garage and vented generators.

9.2- Calculation of Building Surface Areas and Costs.

Later editions of this book will provide computer generated spreadsheets for use by a variety of people from initial planners to architects to user/owners to input data to calculate both window and door areas and all other required information. However, for now, the spreadsheet below will provide a first order estimate with an accuracy of about ± 20% based on a user-friendly simplification.

The building owner usually knows the square feet of living area (under air) plus other areas like the garage. (The garage may contain vehicles, electric and service entrance, power-grid backup transfer switch, electric tools, etc. that must also be shielded to avoid burnout.) User data needed are listed below:

- Number of outside doors

- Number of outside windows

- Average area of each outside window

- Building living ("under air") and working area

- Other area to be protected (for example: garage or basement)

- Average room or floor height

- Length of building

- Width of building

- Number of dormers (homes only, if applicable)

- Number of floors (include basement, if applicable)

- Approximate roof slope in 5-degree increments (homes only)

- Default data: space between floors = 1.5 feet (4' for commercial)

- Default data: commercial roof slope = 0 degrees

- Default data: residential roof slope = 30 degrees

- Default data: residential roof slope, A-frame: 60 degrees

Spreadsheet, Fig. 9.1, provides external shielding areas for new commercial office buildings and EMP-protection cost estimates in percent of building cost.

129

Three building sizes are shown in Fig. 9.1:

Col. B, C, F and G: 50 ft x 100 ft = 5,000 SF, 1 floor, small building

100 ft x 150 ft = 45,000 SF, 3 floor, small-medium

200 ft x 200 ft = 400,000 SF, 10 floor, medium bldg.

A	B	C	D	E	F	G	H	I	J	K	L
Fig. 9.1 – Cost Estimates for EMI Hardened New Commercial Office Buildings											
Topic Identiication	Front (width) Feet	Side (depth) Feet	Num-ber Win-dows	Nbr. Doors	Nbr. Floors	Floor Area Sq. Ft.	Shield Tot Area Sq Ft.	Windows & Doors Sq. Ft.	Remain-ing Sq. Ft.	If Bldg. Cost = $150/SF	emp harden cost in % of bldg cost
Square Feet	50	100	10	3	1	5,000	14,200	303	13,897		
Cost Estimate in $K ($1,000 units)							$88	$5	$83	$750	11.7%
Square Feet	100	150	45	5	3	45,000	97,000	1,185	95,815		
Cost Estimate in $K ($1,000 units)							$593	$18	$575	$6,750	8.8%
Square Feet	200	200	200	7	10	400,000	811,200	4,947	806,253		
Cost Estimate in $K ($1,000 units)							$4,912	$74	$4,838	$60,000	8.2%
Notes: Assumptions: Conductive paint siding = $6/SF and windows & doors = $15/SF at loaded rates											
Windows: 4 ft. x 6 ft. Doors: 3 ft. x 7 ft.											

The building shielding cost for the combination conductive copper paint and aluminum foil siding and for the screened windows and doors are shown in the note in the table at the bottom line. These numbers are based on shielding manufacturer's cost estimates x 3 (for including other components like shielded entrance line filters and surge suppressors, all other direct charges, overhead and profit) to allow for installation and all support.

Column L shows the rough average cost for EMP hardening is about 9% of the new building cost. In early years, like 2013-2014, the cost is likely to be somewhat higher (10-12%). Toward the end of the decade (2017-2019), the cost is likely to approximate 5-6%.

Remember to add the cost of a solar rooftop installation (Chapter 7) because it will become the primary source of backup power for lost electric grid power, and a diesel-engine generator, such as made by Caterpillar, also discussed in Chapter.7.

Although not addressed here, recognize that buildings having more than about three floors may not have sufficient rooftop solar capacity since power is limited to roughly 10 watts/sq. ft. (13 watts/sq.ft. for silicon solar and 8 watts/sq. ft. for thin film) on the roof. For example, in Column G, middle-row in Fig. 9.1, the roof area is approximately 45,000 sq.ft. Use 80%, as a maximum available roof area or 36,000 sq. ft. This suggests that 468 kW is available for a similar building whether three floors or 50 floors. Adequacy of floor capacity break point and other options are beyond the limits of this book now.

But what about EMP hardening if the building **already** exists? The major problem here is difficulty of access to some skin surface areas and or shielding gaps that exist between some covered areas needing bonding of sections. This was discussed in an earlier chapter. Although quite technical and not discussed here, a rough estimate is that the cost may be about 1.3 to double. Perhaps, a 30% default number is suggested as an overall ballpark cost. Adding the solar rooftop and generator (chapter 7), and an EMP hardening for **retrofitting** buildings may approximate 12-15% in the early years of 2013-2014.

9.3 Cost estimates to EMP-Protect Private Homes

Apply the same cost estimate procedure as above to private homes to get the add-on EMP-protection cost. In some homes this may be more complicated than simple rectangular parallelepiped buildings. This is because many/most homes do not have simple rectangular footprints, and have roof pitch angles that can range from $0°$ to $60°$, roof soffit extensions, dormers and the like. However, again, to keep matters simple in this book reading for the lay, some of these add-ons will be included as a percent of the total cost.

Results of the above are illustrated in the table for EMP-protected homes costs in Fig. 9.2. Standby power is added in Column N.

A Topic Identiication	B Front (width) Feet	C Side (depth) Feet	D Nmbr Win-dows	E Nmbr Doors	F Nmbr Floors	G Floor Area Sq. Ft.	H Shield Tot Area Sq Ft.	I Windows & Doors Sq. Ft.	J Remain-ing Sq. Ft.	K If Bldg. Cost = $125/SF	L harden Cost % of bldg	M Roof Slope degrs	N Solar + Batty + Gen'tor	P Total All Costs In %
Square Feet	38	60	15	3	1	2,280	7,231	423	6,808			0		
Cost Estimate in U.S. $							$54,424	$6,768	$47,656	$285,000	19.1%		$26,000	28.2%
Square Feet	38	60	25	4	2	4,560	12,099	684	11,415			30		
Cost Estimate in U.S. $							$90,851	$10,944	$79,907	$570,000	15.9%		$35,000	22.1%
Square Feet	70	40	40	5	3	8,400	19,252	1,065	18,187			20		
Cost Estimate in U.S. $							$144,346	$17,040	$127,306	$1,050,000	13.7%		$44,000	17.9%

Notes: (Assumptions) Conductive paint & Aluminum foil siding @ $7/SF, and windows & doors @ $16/SF installed* add on costs

* Costs are at fully loaded rates which include basic siding/windows/doors, other materials, labor, other direct charges, overhead and profit

Distance between floors = 11 feet

Fig 9.2 is similar to Fig. 9.1, but Column M has been added to allow computing additional increase in rooftop area for roof slope, soffit, etc. Notice, Column L shows an average EMP-hardening cost of 14-19% (more in 2013 and lower, later in the decade).

When the costs of a protected solar rooftop, battery and generator back up power are added, Column N results. When Columns H (EMP-hardening) and Column N (backup power) are added, the total of all estimated costs are listed in Column P. This is shown in terms of per cent costs relative to a new home costs. Again, these costs will be higher in 2013 and lower in 2020. As mentioned earlier, The costs are also higher for retrofit of an existing home.

Chapter 10

Who Pays for EMP Protection and How?

Chapter 10 Overview

Little will happen until EMP financing options are identified and the supporting mechanisms are funded and placed into action. Of course, before funding EMP hardening in the first place, other higher priorities must be first identified and planned and many convinced that *EMP insurance is really needed. (Note:* County bond financing as discussed in chapter 2. is totally different than financing in this chapter since the former is based on a Four-tier EMP protection plan concept for the entire U.S.)

The matter of solar rooftop (no EMP involved) funding is entirely different and already has substantial exposure, with rebates, tax credits and feed-in tariffs. These are the most substantial ranging from 30% to 60% of the installation cost in the U.S. They are expiring.

In early 2012, Germany cut back its Feed-in-Tariffs; then more in April and dropped back by 2%/month thereafter until further notice. Other countries, including Italy, are considering cutting rebates or have already done so. This applies to many U.S. states as well. Remember Germany and Italy constituted about half the world solar market in 2010; far less thereafter, as China has since become the world's largest producer..

State subsidies of 30% have expired. The U.S. federal tax credit for solar-PV expires in 2015. Bottom line is that manufacturers have a glut of inventory; prices have fallen 45% in one year, and many companies have sustained big financial losses and some have gone bankrupt. However, this is forecast to gradually end in about 18 months. More than 90% of all solar sales in the U.S. today are for solar farms.

The above goes on while economic expansion is seen in the average *employment multiplier.* This is an economic term referring to the multiplication effect that a region, state or county can have in its jobs via exporting its products and services.

10.1- EMP Protection, Potential Financing Options

The inclination to continue to ignore the EMP threat or to start planning or initiate action, will be heavily dependent on:

(1) Who is to pay? Where is the money coming from? What kind of funding is available or might be available, and when?

(2) Is the home or building already built or is it to be built? In either case, is the structure to be EMP hardened? Is solar electricity to be added? When?

(3) What is the degree of world tension? What is the latest nuclear threat from Iran and North Korea? What blunders has the U.S. made lately to irritate the Taliban and al Qaeda radicals (like burning the Quran)?

(4) Because of the results of the U.S. election in November 2012, more government intervention into the economy is expected. So, the financing explained in this chapter is more to be expected than the more free market approach scenario outlined in chapter 2.

These and others all impact in varying degrees the perception of "EMP", "EMP protection" and the sources of financing.

Since the hardest part of getting started is *getting started*, look at Fig. 10.1. It illustrates what is involved in attempting to optimize an enormously complex structure such as the U.S. infrastructure. Where to begin? What and where to optimize?

Select two groups in Fig 10.1 on the next page for the moment to help start: (1) The electric grid since that is where 98% of the electricity originates and (2) Homes of the wealthy. The latter is the easiest since the wealthy can readily spend, say, $200,000 in a heartbeat to EMP-protect both their primary and a get-away home and its solar rooftop installation. This basically may be the situation described in chapter 7, *Off-the-Grid, EMP protection.*

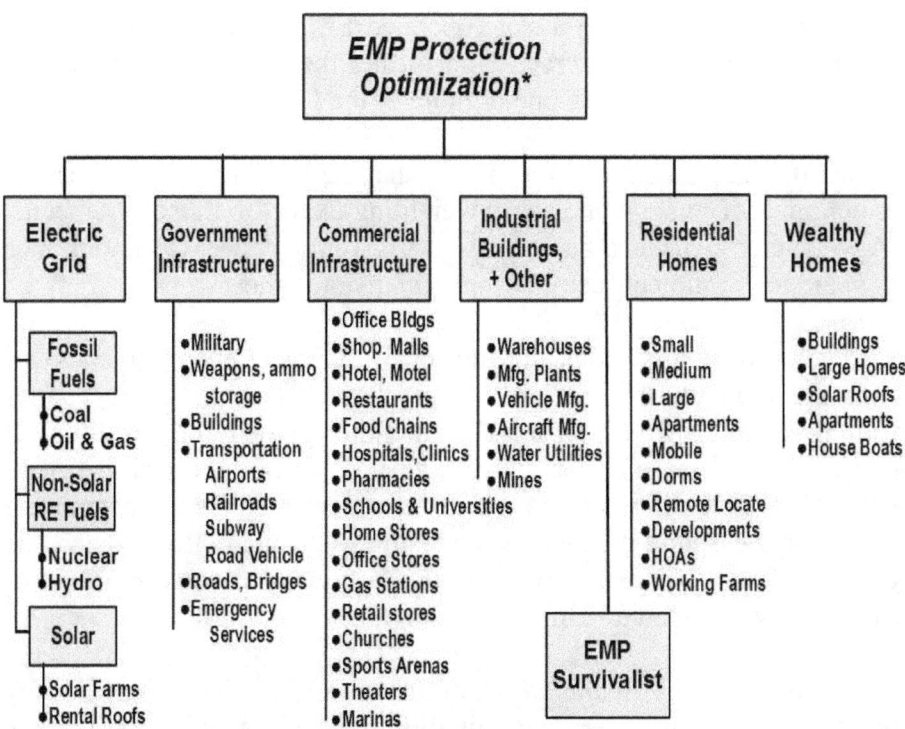

Fig. 10.1: Six major components used in EMP Optimization

This then leaves #1 above, *the Electric Grid*. Since that is the subject of Vol. 4 of this EMP handbook series, it will be deferred until then. So pick one more for now. Let's select Commercial Infrastructure since if one big sector fails, then millions of people suffer without food or other vital survival supplies. In other words, how do we finance EMP hardening for *commercial structures*? Another reason for selecting this now, is that it will likely be less difficult since the commercial world already has trillions of dollars in resources and it is the biggest (including industrial) of the electric utility loads (more than residential).

These considerations are why at the beginning of this handbook with respect to the status of EMP, it was said that a cataclysmic EMP incident is "Not If, but When"? Thus, two general approaches for generating financing are candidates now:

(A) From the bottom up: So the assumption for this section is that the *Economic Development Offices* of the 3,140 counties in the U.S. will have a mandate to start thinking and doing some early EMP awareness, reflections and planning. Their discoveries, observations, inclinations, creativity will flow up to a state operated EMP Action Group (EAG). The state will examine, screen and further the best of the approaches offered which may include one or more of the following:

• Identify the parameters of what it is that the EMP hardening is trying to optimize? Generate and apply weighting factors to those parameters. (A weighting factor is a numerical score given to reflect the importance or value of the subject sector or topic relative to others).

• Develop or contract out and run local EMP awareness slide shows, with the thought of "crowdsourcing" the reactions and suggestions from the slide show participants. Be sure to have the more vocal (or all) participants fill out and turn in an evaluation form or record the meetings. Crowdsourcing is also a powerful way to identify the parameters and their weighting factors to make things happen. Read the book, *Crowdsourcing – Why the Power of the Crowd is Driving the Future of Business* by Jeff Howe. This is a must read.

(B) In the U.S., the *Economic Development Offices* have the fiduciary responsibility of discovery and initiating (with ideas and help from the Chamber of Commerce) future County financial health and growth prospects and follow through of *Employment Multiplier* concepts as outlined next.

10.2 Employment Multiplier

Employment multiplier (EM) is an economic term referring to the multiplication effect that a region can have in its jobs via exporting its products and services. While it takes X employees to manufacture and ship its products outside the region at Y_0 dollars, the income, Y_i, it gets is far greater than Y_0. This produces excess wealth in the region to increase salaries and/or generate more jobs. Since these employees spend most of their income locally, this generates additional jobs for people, who spend most of their income locally, etc.

One way to think of community economic analysis is to imagine the community's economy as a barrel with money and goods flowing into the top as well as spilling out (Fig. 10.2). The barrel analogy represents a number of key concepts. First, the community is intimately linked with the rest of the world through the inflow and outflow of income and goods. Second, the community uses resources to produce the output it sells. These resources can be available locally or purchased elsewhere. Third, the size of the barrel is determined essentially by the inflow of outside income, the lack of leakage of income, and the volume of resources used to produce the community's output.

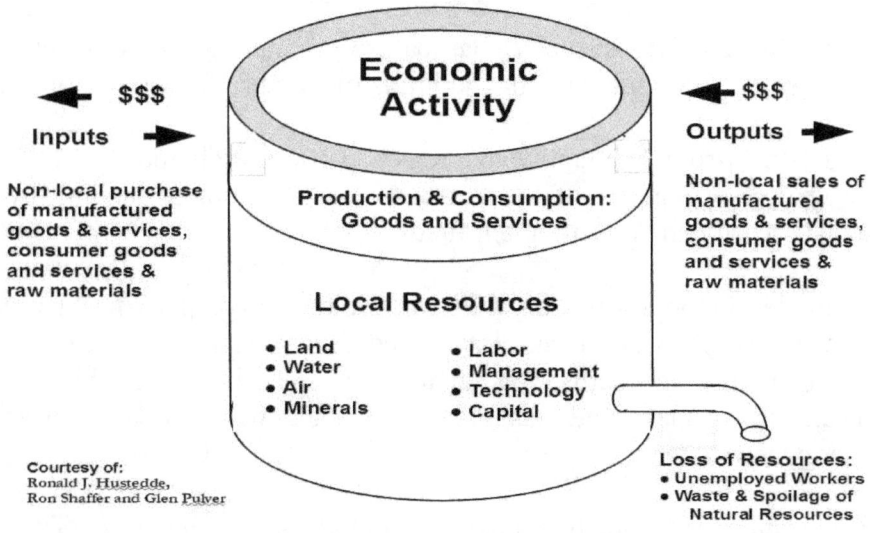

Fig. 10.2 – Illustrating the Employment Multiplier Concept

The region mentioned above can be a city, county, or municipality. The region may have a positive or negative export plurality over imports. If positive, the region will grow in wealth and jobs, and vice versa. A classical example at the country level is China and the U.S. The United States use to be a world exporter with an EM ratio of $Export/$Import much greater than 1.0 for many decades and grew globally in wealth and jobs accordingly.

In the past few decades, however, the U.S. has imported more than exported, especially from China. So today China owns more than a trillion dollars in U.S. Treasury notes and grows at 6-10% annually in employment while U.S. stands still and is struggling in a recession. The reason offered by some is very simple. The labor unions ran up their salaries beyond competitive rates. So manufacturers went offshore to get their products made at lower rates. Result: EM is less than 1.00 for the U.S. today. Who is watching and reporting our store?

The four biggest contributions to a regional EM are:

(1) Manufacturing (or a high tech or other outside-needed service not readily found elsewhere): which sells mostly to outside its region .

(2) Tourism: which brings the visitors and their money to the region.

(3) Seasonal activities: such as baseball spring training which selects a southern U.S. city for stadium and infrastructure use for a few months.

(4) Retired people who move to a region. They bring in money and jobs by way of a plurality of visitors to a vacation spot or income from social security and/or retirement funds.

Example: Suppose a retiring CEO of a mid size company has a vacation spot on the Chesapeake Bay called Tiffany Harbor. His company, headquartered in metro Washington, D.C., is in the banking and insurance business and has close associates who run a renewable energy newsletter.

Three enterprising neighbors have week-end vacation homes on Tiffany Harbor on the Chesapeake Bay, 95 miles from metro Washington. They notice that while their development seems to be somewhat exclusive, the rest of the county is devoid of resources, such as food stores, gasoline stations, retail outlets, restaurants, clinics, etc.

Since the round trip to their homes in metro Washington, D.C. takes 3-1/2 hours every weekend, they need some Tiffany Harbor nearby commercial stores so that they can vacation for a week from time-to-time and support emergencies. The three are in the banking, magazine and renewable energy business. They envision renting or buying a nearby house and setting up the makings of a three-business extension offices in the house. They have access to all the local Tiffany Harbor help needed for staffing from their own waterfront community.

Their plan is to primarily broker their skills from Tiffany Harbor in behalf of their present employers:

(1) The bank provides financing services (or referrals) for Tiffany Harbor residents (refi, reverse mortgage, solar installations, new residential and commercial property sales).

(2) The magazine services a trade organization, but all the marketing work by one of the three is done over the phone, cell, Skype and iPad – no customer visits. The first hand exposure to events at Tiffany Harbor make for some excellent magazine short stories.

(3) The renewable energy guy is an editor who can do his work anywhere and provide solar knowledge to the other two as needed. Their drafted three-year business plan shows that their combined office will:

(a) Gross about $4M in behalf of their employers and over $1M for extra brokering for the benefit of Tiffany.

(b) In addition to banking, broker insurance is offered for local homes and automobiles, solar installations and EMP hardening, as it unfolds.

(c) Reduce the need for and savings on their home office space and associated overhead.

(d) result in purchase of four Tiffany Harbor homes by associates back at Washington, DC in the first two years.

(e) Addition of a nearby gasoline station with food store, a pharmacy, an overdue clinic to service Tiffany Harbor, a small volunteer emergency service station and others.

(f) The threesome represent a significant input to the county's Economic Development office, county administrator and board of supervisors.

The new taxable revenue base generated (IRS, state and county) from the above represents approximately $3.5M in the first 18 months.

FYI: every new job created in USA is worth roughly $40,000/year to the recipient, and for every $90,000 spent into the community, one new job is generated (cf: *"A Blueprint to Low Unemployment"*, by Don White).

The above story was fashioned after a real life experience by changing some of the names. The point is that there is a myriad of other versions similar to the above that constitutes how EMP can be financed from the bottom up, in addition to what the government may do from the top down.

> *Employment Multiplier* can add 10% to 30% more jobs for most of the upper 4-Tier Counties. (See chapter 2.)

10.3 Solar Rebates, Tax Credits & Other Financial Incentives

The previous section provides part of the backdrop for EMP protection of all manner of buildings because solar is expected to be on par with fossil fuel replacements (coal, oil and gas) for generating electricity by 2017-2020. Because solar is only available at daytime and with minimum cloud cover, the solar backup of batteries and generators are discussed in chapter 7 in off-the-grid living. One of the benefits in southeast California (for example, the Mojave Desert) of the large windmill farms is that they work through the night when solar is inoperative.

The next section provides a number of specifics on solar financing some of which can be adopted to EMP financing as applicable. Meanwhile a few words about solar rebates and tax credits is relevant. They have been very instrumental between 2006-2011 in solar rooftop installations in that they financed between 30% (rebate only) and 60% (rebate and tax credit) of each contract in the U.S. Since solar-PV was about seven times the fuel bill in electricity development in 2007-2009, they still resulted in a cost of 3-5 times oil/gas/coal.

However, since the 2011 collapse of solar panel pricing due to overproduction when Germany announced its rebate will be withdrawn, solar is about 2.5 times their fossil-fuel counterpart in 2012. With Moore's Law at work, par is expected to be reached in about five years, rendering the value of rebates and tax credits of questionable value today. Even First Solar announced it reached a $0.78/watt cost for producing CdTe solar panels in 2011.

10.4 Solar Financing Options and Other Derivatives

Fig. 10.3 illustrates many different financing options that might be available in any given solar installation. Equity financing (left block, friends and family) may be the only route available to a homeowner who has less than a good financial score (for example, less than about 700). In that case the homeowner is likely to have too small a home (less than, say, 1,800 sq. ft. in 2010. However, very soon, Moore's Law will likely result in a much lower (45-60%) solar installation cost.

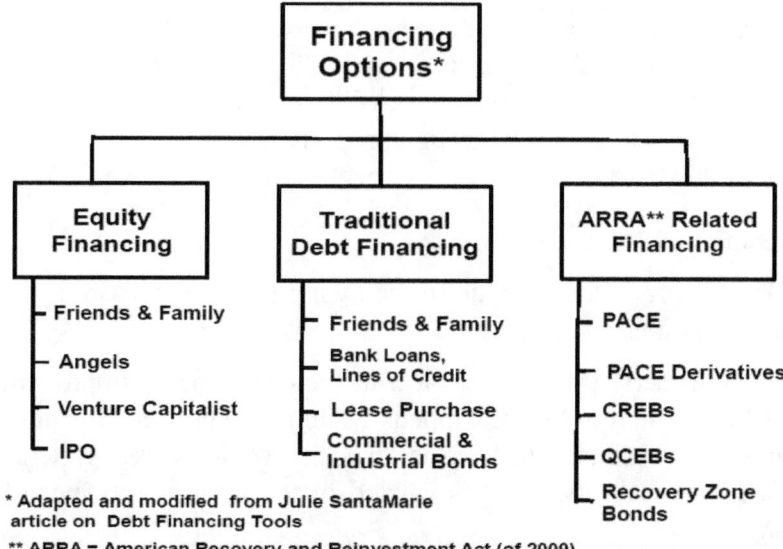

Fig. 10.3. Multiple financing options for solar projects and programs.

For commercial office and other non-residential buildings, solar *Equity Financing* may be a part of a much larger construction situation. This will not be discussed further in this handbook as that emphasis goes well beyond solar installation financing.

Traditional Debt Financing is shown as the second block in Fig. 10.3. Historically, this has been a major part of the slowness in residential solar rooftop installations in the past. Even with state rebates and federal tax credits, the median U.S. homeowner (2,300 sq. ft. under air) would have to come up with $10k to $20k in cash as a down payment or otherwise finance this. Because breakeven (discussed in Section 4.0) may run from 10-15 years (or more), this was typically a deal breaker, further exacerbated by the fact that the installation typically (then 5 kW) covered only about half of the house electrical bill.

Berkeley, CA FIRST

The Berkeley FIRST program served as a model for Property Assessed Clean Energy (PACE) programs across the country. The pilot program provided property owners an opportunity to borrow money and repay the financing through their own property tax bills over 20 years.

PACE Financing

Basically, PACE is a way to finance solar system installations or energy efficiency retrofits, where a city, municipality, county or state provides a loan developed from bond sales. As previously stated, the home or building owner pays it back through property tax additive over a period, usually of 20 years.

The benefits of PACE are:

• The program doesn't require the owner to pay any cash (or small upfront amount at settlement).

• Using property tax increases to finance clean energy improvements solves the problem of "what happens when an owner sells his house or building?" The simple answer is that the solar power system and associated tax liability (lien) is carried over to the new owner of the building.

• PACE is backed by the bond-issuing authority so participants know they won't get scammed, and the financing is guaranteed to be secure.

An announcement in 2010 by the FHFA (Federal Housing Finance Agency) has declared that residential PACE financing programs do not meet the financial requirements of federal mortgage giants Fanny May and Freddie Mac. This announcement significantly hampers the ability of homeowners to use PACE financing to fund clean energy and energy efficiency projects for their home. It also effectively halts new PACE program development in the near future as PACE supporters are forced to appeal to elected government officials to intervene on PACE's behalf. At the time, it was expected that the FHFA problem will be resolved on or by spring 2011. Unfortunately, this did not happen even though the benefits far outweigh the petty politics.

Derivatives of PACE Financing

While the substantial price reduction of second generation, thin-film, solar rooftop installations make "going solar" very compelling in most USA states, and further compelling by new financing, it may not work for all situations. Some legislation may be needed at the County or other regional level to ensure that laggards, procrastinators and disbelievers, are "further motivated".

One example (listed in the upper right box of Fig. 10.4) from the past, is the financial viability of putting in central water distribution when many or most existing homes were already on wells. Many may not want to convert from a well to central water distribution. So, what does a city do?

A central water conversion moratorium was typically set up, whereby building owners (usually homeowners – not commercial buildings) are given one to three (or more) years to hook up or otherwise be assessed (impact fee) for the conversion, anyway.

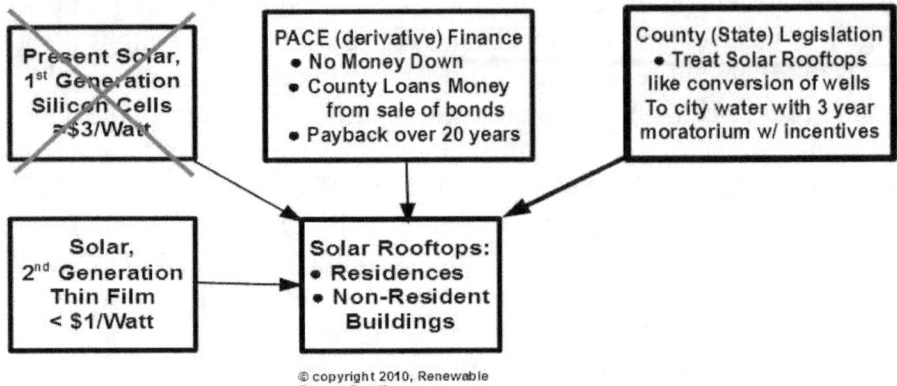

Fig. 10.4 – One derivative of PACE involving legislation to effect a solar rooftop installation requirement.

To the extent applicable, this or a related process may be needed for solar also by some locations or jurisdictions. In this way, then, the vast majority of detached homes and commercial and industrial buildings can be counted on for further price reduction by the benefit offered through quantity purchases.

Note: with a low price of thin film, solar installation (< $1.00/watt by mid or late 2010s), the financial benefits of "going solar" become so compelling that a moratorium may not even be needed.

Seniority of Mortgage Loans

Another problem with PACE in 2010 is the seniority of the PACE-derivative loan. Most PACE backers want the loan to be made senior to the first mortgage loan. This mitigates the problem of foreclosures and reduces the risk of "short sales" - two topics that are important to bond buyers of the PACE program. The counter argument by banks and lenders, is that it will compromise the security of their mortgage loans.

One solution of many, may be to treat the lien placed on the home like a mechanics lien, although not to be paid off at the settlement of a sale. However, the condition for the most senior PACE position is limited to, say, for example, the first $5,000 only. The remainder of the loan is junior to the first mortgage holder, but senior to any other loans.

As you can see from the above, there are many combinations of possibilities to provide a satisfaction to the seniority of the PACE-derivative loan. Therefore, this should not become a major problem.

Another aspect of the PACE-related program is what homes get covered with rooftop solar and what are the selection criteria? Obviously, not all detached and duplex/villa homes can/should be included for PACE derivative financing. How to deal with roof area size of a house, Moore's Law, roof ages, and financial responsibility, etc. One logical approach in the starting years is to require that the new standards and specifications have a, say, three-year life, whence it must get up-dated before expiration, (continues on by default but with annual updates).

Recognize in the near term, it is not desired that all detached and duplex homes be included. For example, if the lower third in value and/or roof area were eliminated, the median electricity monthly bill will increase by roughly 50% as discussed in Chapter 7, *"A Blueprint to Low Unemployment"* by Don White. This greatly increases the median "useful" money to be saved and then spent later back into the economy to generate more jobs.

Another consideration involves HOAs (Home-Owners Associations). It is not unusual for some HOAs to take a somewhat negative view of solar rooftop installations. They may be regarded as unaesthetic or a threat to the power or sovereignty of the HOA board of directors or their architectural approval committees. In extreme situations, they may be cited in the deed restrictions.

The Community Institute Trade Association estimated that HOAs governed 23 million American homes and 57 million residents in 2006. This corresponds to 19% of all residential homes and people in USA. It is believed that most would not oppose solar rooftop installations. This leads to less than 10% of available rooftops posing a potential problem. However, there exists many solutions including giving the "recal-citrants" a say, five-year moratorium to join up or lose the PACE derivative or other financing benefits of failing to join. Before the end of this decade, when they see what they are missing out on, it is expected that few would object while still specifying some references in their deed restrictions.

Then, who gets included at the start? Since the median size (2,300 sq. ft. under air) and larger homes account for 85% of the electric bill in an average residential region, they would qualify if their roof-life expectancy has at least, say, 15 remaining years. If lesser life, they receive a rain check or the financial loan duration is correspondingly reduced.

Feed-in Tariffs

A feed-in tariff is a policy mechanism designed to accelerate investment in renewable energy technologies. This is achieved by offering long-term contracts to renewable energy producers (sometimes users), typically based on the cost of generation of each technology. Technologies such as wind power, for instance, are awarded a lower per kWh price, while technologies such as *solar-PV* and *tidal power* are offered a higher price, reflecting higher costs.

Feed-in Tariffs are also known as *FITs*. They are the electricity part of what some people call *Clean Energy Cashback*. To some, FITs are a government scheme that pays people for creating their own *green electricity. In UK, for example,* Tariffs give three financial benefits:

- A payment for all the electricity produced, even if the owner uses it himself .

- Additional bonus payments for electricity the owner exports back into the electric grid.

- A reduction on the owners standard electricity bill from using energy produced himself.

FITs are for everyone, including households, landlords, businesses and even organizations such as schools and care homes.

Additionally, feed-in tariffs often include "tariff digression", a mechanism according to which the price (or tariff) ratchets down over time. This is done in order to track and encourage technological cost reductions (see Moore's Law). The goal of feed-in tariffs is to offer cost-based compensation to renewable energy producers, providing price certainty and long-term contracts that help finance renewable energy investments.

10.5 The EMP Insurance Viewpoint and Options

The developed nations of the world are accustomed to purchasing insurance as a protection to reduce the risks of catastrophes involving loss of human life, homes, buildings, automobiles, jewelry, antiques, expensive collectibles, travel, events, and much more. Why not take out insurance against an EMP event, or better yet, for an EMP hardening of a home, commercial office building, shopping center, hotel/motel, church, business, etc?

With a bit of poetic license, insurance can also take on the form of other than paying a traditional periodic payment premium to offset the risk of loss of the protected item or structure. For example, add up what you would pay in premiums over, say, a 20-year period on (1) adding a solar-PV rooftop for free electricity or (2) to have your home EMP protected. Here are two examples for further thought:

(1). Assume your monthly electric bill, averaged over a year is $175/month = $2,100/year = $42,000 for 20 years. (For simplicity, no adjustment is made here for inflation). The cost for an average 7 kW power solar rooftop installation in mid 2012 at $2.75/watt = $19,250. If 100% is financed at 5% per annum interest, the total cost is $30,491. Added insurance and maintenance for this is about $4,000, totaling roughly $34,000. If you move before 20 years, the remaining solar installation unpaid amount carries over to the new owner. Thus, you saved $42,000 - $34,000 = $8,000 and you have free electricity and a solar rooftop installation along with its bragging rights and home improvement value increase. In this case, you insured yourself since you pay no insurance premiums, but pay the bank or lending source for the solar installation. Perhaps, best of all, your break even and positive ROI begin as soon as the solar installation is operational (less than one month).

(2) Case #2, is the cost to EMP-protect your home, knowing that you can survive an incident if it happens even though all your neighbors may not be so protected. Further suppose, that the cost to EMP-protect your home is about 8% for an existing home and 5% for a new home. If the market value of your existing home is $300,000, you can finance the 8% or $24,000 similar to the above example. In this case assume that the resulting $158/month payment to the bank is an insurance premium for survival in the hopefully unlikely event of this holocaust.

Case #2 can be further enhanced by combining it with case #1, so that if an EMP incident should occur, you make your own electricity and life goes on. All the neighbors, not so equipped, would look to you for survival. Maybe you need to add a refrigerator or two in your garage as standby and perhaps other appliances for the neighbors who could pay into this mini insurance.

Of course this could lead to your own homeowner's association (HOA) to dedicate a small building, EMP protected and with solar rooftop, battery and generator, a variation of what you see in Chapter 7. Then, all participants would have their own or share partial refrigerator, air conditioner, hot water, dishwasher and dryer and lighting. If you had 20 HOA members paying $40/month for this insurance, that's almost $10,000/year for buying a 3,000 sq. foot building and equipping it with survival outfitting for all.

Index

A

B

C

D

E

F

G

H

I

J

K

L

M

N

O

P

Q

R

Racking, Solar Panels, 88
Radioflash, 21
Rainwater harvesting, 164
Recovery Zone Bonds,
References, 22, 31, 36
Renewable energy, 78, 79, 103, 138, 19, 146, 158, 159, 161
Renewable Energy Creations, LLC, 3, 156
Renewable energy financing
Return on Investment for solar rooftop installation,
R-insulation ratings, 172
Rockets, Redstone, Thor, 25
ROI = Return on Investment, 13, 44, 90, 91, 95, 147, 164
Roof slope, solar, 106

S

Santa Marie, July, 141
Savings spent back into economy
SCADA = Supervisory Control and Data Acquisition Systems, 19, 20
SCADA, shield critical components, 20
SGEMP = System Generated EMP, 166
Shell Company, 167
Shielded building entrance(sm) 97
Shielded Buildings, Rooms & Cabinets, 64
Shielded inverters and micro-inverters, 99
Shielded panels, bonding, 97
Solar rooftop victims,
Solar site visit & analysis, 87
Solar technology, 2nd generation
Shielded system wiring, solar farms, 126
Shielded windows, doors and other, 73
Shielded windows, wire mesh, 73
Shielded windows, copper wire screening, 74
Shielded wires and cable entries, 75
Shielding of the Building Facade, 66
Shielding conferences, 41
Shielding effectiveness, screen equation, 74
Shielding guidelines
Shielding and grounding solar trackers, 126
Shielding, How it works, 64

Solar shielding effectiveness requirements, 60
Silicon, amorphous, 82
Silicone, crystalline, 82
Slide show, Jobs-Solar-Pace- EM for teachers/instructors
Smart grid, 167
Smith, Gary, 17
Soil resistivity, measurements of, 63, 71
Solar, break even, 91-95,
Solar cell costs
Solar cell efficiency,
Solar cell technology,
Solar cell, warranty
Solar cost, 79, 92
Solar cost per watt, 92
solar earth seasons, 167
Solar efficiency and costs,
Solar energy perception,
solar farms, 120-123, 125, 128
Solar farms, Arcadia, Babcock, China, 121
Solar farms, representative examples, 121
Solar farms, non-photovoltaic, 122
Solar farms, parabolic trough, 122
Solar financing options, 129, 140, 147
Solar flares, 2, 15, 18, 163, 167
Solar foundation Industry,
Solar 1st generation, 81, 162
Solar 2nd generation, 81, 162
Solar 3rd generation, 81, 162
Solar generations, 81
Solar generation and PACE Financing,
Solar installation costs, 103
Solar installation outlives roof,
Solar inverters & micro-inverters, 90
Solar jobs, average pay,
Solar jobs score,
Solar jobs, unemployment reduction,
Solar PACE financing, 142-143
Solar PACE financing, derivatives, 141
Solar paints, inks and dyes,
Solar-PV panels, 68, 76, 80, 84, 97, 105, 107, 124, 141, 159, 179
Solar panels, shielded, 96
Solar panel racking, 89
Solar peak power, 82, 83

T

U

V

W

X-Y-Z

Manufacturers of Source Materials and/or Serices which may be used for HEMP & Solar Hardening

#	Company Name	Tel. Number	Fax Number	E-mail	URL-Website	SR	SC	SW	SH	BD	GR	SS	FT	TL	CS
1	Chomerics	781-935-4850	781-933-4318	Chomeribox @											
2	ETS-Lindgren, Illinois	630-307-7200	630-307-7571	info @ets-lindgren.com	www .ets-lindgren.com			■	■	■	■			■	■
3	ETS-Lindgren, Texas	512-531-6400	512-531-6500	info @ets-lindgren.com	www .ets-lindgren.com					■	■			■	■
4	Holland Shielding	31-78-613-1366	31-78-614-9585	info @hollandshielding.com	www .hollandshielding.com										
5	Metatech Corporation	805-683-5681	805-683-3023	info @metatech.com	www .metatechcorp.com	■	■								
6	PanaShield	203-866-5888	203-866-6162	help @panashield.com	www .panashield.com		■								
7	Protection Technology Group	800-882-9110		sales@protectiongroup.com	www .protectiongroup.com						■				
8	Renewable Energy Creations	941-743-7633	941-743-7633	drjw9@aol.com	www .emp-protection.org							■			
9	Retlif Testing Laboratories	703 533-1614	703 533-1612		www .Retlif.com										
10	RFI Corporation	631-231-6400	631-231-6435	rfisales@rficorp	www .rficorp.com										
11	Schaffner		732-225-4789	ken.bellero @schaffner.com	www .schaffnerusa.com										
12	Schurter	707-636-3000	707-636-3033	info @shurter.com	www .schurerinc.com				■						
13	Spectrum ASP	814-474-1571	814-474-3110	susan @altman-hall.com	www .spectrum.com								■		
14	Tech-Etch	548-747-0300	548-746-9639	bmcallister @tech-etch.com	www .Tech-etch.com				■						
15	Zippertubing	800-321-8178	310-767-1714	orders @zippertubing.com	www .zippertubing.com		■								

A Few Solar Panel & Inverter Manufacturers and Solar Magazines

#	Company Name	Tel. Number	Fax Number	E-mail	URL-Website	SR	SC	SW	SH	BD	GR	SS	FT	TL	CS
						Panels		Inverter	b Magaz.						
16	First Solar	877-850-3757	602-414-9400		www .firstsolar.com										
17	Suntech Power	866-966-6555	415-882-9923		www .suntech-power.com										
18	Sun Power Corp.	800-786-7693			www .us.sunpowercorp.com										
19	Yingli, China/USA	888-686-8820	212-686-8806	info @yingliamerica	www .yinglisolar.com										
20	Canadian Solar	925-866-2700	925-866-2704	inquire.us @canadian	www . Canadian-solar.com										
21	Solectria Renewables	978-683-9700	978-683-9702	inverters @solren.com	www .solren.com										
22	Delta Products Corp	877-460-5851		sales.usa @solar-inverter.com	www .solar-inverter.com/na										
23	Global Solar Technology	239-245-9264	239-236-4682	arae@globalsolartechnology.com	Www .globalsolartechnology.com										
24	Solar Industry	800-325-6745	203-262-4680	info @solarindustry.com	www .solarindustry.com										
25	Solar Today	303-443-3130	303-443-3212	ases@ases.org	www .ases.org										
26	Photon.Info	617-874-5500	617-262-4309	info @photoninfo.con	www . Photon.info.con										
#	Company Name	Tel. Number	Fax Number	E-mail	URL-Website	Panels		Inverter	b Magaz.						

Code: SR = shielded rooms, enclosures. SC = Shielded cables, connectors. SW = shielded windows, glass panels. SH = shielded materials, windows, building siding, other
BD = bonding. GR = grounding related products and subjects. SS = surge suppressors. TL = Testing Labs, CS = consulting
Note: we will be expanding this list in our next printing. If you wish to be listed, please provide your credentials

Curriculum Vita of the Principal Author

Don White, registered professional engineer, ret'd., holds BSEE and MSEE degrees from the University of Maryland. He is past CEO of three Electromagnetic Interference and Compatibility companies in Metro Washington, D.C., one of which traded on the NASDAQ.

Don has written and published 13 technical books over a span of 30 years, which became well known and used worldwide in electronics circles. His last book was "The EMC, Telecom and Computer Encyclopedia Handbook", an 800-page compendium. Don taught over 14,000 engineers via seminars in 39 countries.

At Don White Consultants, he published a bimonthly trade journal called *EMC Technology* magazine circulated over four continents. In addition to being technical editor, he wrote many of the tutorial articles.

Don received the IEEE award for development of automatic spectrum scanning, recording and analysis intercept systems. A 2nd award for contributions to the EMC education and publishing arena. He is a senior member of the former Institute of Radio Engineers and life-time senior member of IEEE. He is past president of IEEE, EMC Society.

Don can be reached at www.emp-safeguard.com or 941-743-8100, or drjw9@aol.com. See page 183 for information on updating different printings of this 3rd edition and the planned 4th edition to include advertising, presentations and videos.

Other books by Don White aka Donald R.J. White

- **Nuclear EMP Threats – What Next?,** REC Press, 2012, 162 pages,
- **Save, Survive and Prosper in an Economy in Crisis**. Dougherty and White, 132 pages, 2008, D-W Press
- **Handbook of Electromagnetic Compatibility,** White and Violette, 2002, Van Nostrand Reinhold
- **The EMF Controversy and Reducing Exposure from Magnetic Fields,** White, Barge, George and Riley, 201 pages, the EEC Press
- **The EMC, Telecom and Computer Encyclopedia, Third Edition,** 800 pages, 1999, *emf-emi control Press*
- **12-Volume Handbook Series,** 1988, The EMC Press:
  - **Vol. 2, Grounding and Bonding,** 487 pages
  - **Vol. 4, Electromagnetic Shielding,** 615, pages
  - **Vol. 6, EMI Test Methodology and Procedures,** 675 pages
  - **Vol. 8, EMI Control Methodology and Procedure,** 544 pages
- **Shielding Design Methodology and Procedures,** 1987, DWCI Press
- **EMI control in the design of printed circuit boards and backplanes** 1981, DWCI Press
- **A handbook on electromagnetic shielding materials & performance** 1975, DWCI Press
- **Five-Volume handbook Series on Electromagnetic Interference and Compatibility,** 1972, DWCI Press
- **Glossary of Acronyms, Abbreviations & Symbols,** 1971, DWCI, Press
- **Methods and Procedures for Automating RFI/EMI Measurement,** 1966, WEI Press
- **Electrical Filters – Synthesis, Design & Applications,** 295 pages, WEI Press

Curriculum Vita of the Co-Author

Jerry Emanuelson received his degree in Electrical Engineering from the University of Colorado. He began his career as a test engineer for Ampex Corporation, where the video tape recorder was invented. There he wrote test procedures and designed test equipment for audio and video recorder circuitry.

He later became the Transmitter Supervisor for a broadcasting company with transmitter locations on mountaintops subject to severe lightning and significant electromagnetic interference problems due to many types of closely packed transmitters.

He is a former member of the IEEE Broadcast Society. He is now a part-time electronics consultant and a part-time science writer.

For many years, he has maintained a web site that includes several pages on the subject of nuclear electromagnetic pulse. He has been thinking about the EMP problem on civilian infrastructure for more than three decades. He joined Don White as a co-author of the 3rd edition of this first volume of the EMP series.

His futurescience.com web site will have a frequently updated Appendix D to this book. It will cover some of the topics that may need updating too frequently to be appropriate for inclusion in this book. This will include such topics as getting from the individual level of protection up to the community level. It will also include things like hints about how individuals can EMP-protect new solar panel installations. Look for this new Appendix D at:

http://www.futurescience.com/emp/AppendixD.html

Appendix A. Terms and Definitions

The following lists some of the most often appearing terms in EMP and Solar installations and their protection.

Alternating Current (AC): The flow of electricity that constantly changes direction between positive and negative sides. Almost all power produced by electric utilities in the United States moves in current that shifts direction at a rate of 60 times per second.

Alternative Energy: Another name for renewable energy – alternative to fossil fuel (oil, gas and coal).

Ambient Temperature: The temperature of the surrounding area.

Amorphous Silicon: A thin-film, silicon photovoltaic cell having no crystalline structure. Manufactured by depositing layers of doped silicon on a substrate. See also single-crystal silicon an polycrystalline silicon.

Ampere (Amp): The unit of measure that indicates how much elec-tricity flows through a conductor. It is like using cubic feet per second to measure the flow of water. For example, a 1,200-watt, 120-volt hair dryer pulls 10 amperes of electric current (amps = watts/volts).

Ampere-Hour (Ah/AH): A measure of the flow of current (in amperes) over one hour; used to measure battery capacity.

Annual Solar Savings: The annual solar savings of a solar building is the energy savings attributable to a solar feature relative to the energy requirements of a non-solar building.

Average Demand: The energy demand for a given location over a period of time. For example, the number of kilowatt-hours used in a 24-hour period, divided by 24 hours, tells the average demand for that location in that time.

Avoided Cost: The amount of money an electric utility would need to spend for the next increment of electric generation to produce or purchase.

Azimuth Angle: The angle between true south and the point on the horizon directly below the sun.

Battery: Batteries are often sold with a solar electric system. The primary purpose is to store the electricity not immediately used, which could be used at some later time.

Billion: = 1,000,000,000 = 10^9 = 1,000 million

Biomass: a renewable energy source, biological material derived from living, or recently living, organisms, such as wood, waste, and alcohol fuels. Biomass is commonly plant matter grown to generate electricity or produce heat.

BIPV (Building-Integrated Photovoltaics): A term for the design and integration of photovoltaic (PV) technology into the building envelope, typically replacing conventional building materials. This integration may be in vertical facades, replacing view glass, spandrel glass, or other facade material; into semitransparent skylight systems; into roofing systems, replacing traditional roofing materials; into shading "eye-brows" over windows; or other building envelope systems.

BTU (British Thermal Unit): The amount of heat required to raise the temperature of one pound of water one degree Fahrenheit; equal to 252 calories.

Cap and Trade: A central authority (usually a government) sets a limit or cap on the amount of a pollutant that can be emitted. Companies or other groups are issued emission permits and are required to hold an equivalent number of allowances (or credits) which represent the right to emit a specific amount. The total amount of allowances and credits cannot exceed the cap, limiting total emissions to that level.

Companies that need to increase their emission allowance must buy credits from those who pollute less. The transfer of allowances is referred to as a trade. In effect, the buyer is paying a charge for polluting, while the seller is being rewarded for having reduced emissions by more than was needed. So, in theory, those who can easily reduce emissions (such as renewable energy producers) most cheaply will do so, achieving the pollution reduction at the lowest possible cost to society. The "Cap and Trade" is believed by many to create more damage than good.

Capacity Factor: The ratio of the average load on (or power output of) an electricity generating unit or system to the capacity rating of the unit or system over a specified period of time.

CIGS = copper, indium, gallium, and selenium, thin film, solar semi-conductor used for 2nd generation solar panels

Circuit: One or more conductors through which electricity flows.

Concentrator: A photovoltaic module, which includes optical compo-nents such as lenses (Fresnel lens) to direct and concentrate sunlight onto a solar cell of smaller area. Most concentrator arrays must directly face or track the sun. They can increase the power flux of sunlight hundreds of times.

Converter: An electrical apparatus that changes the quantity or quality of electrical energy.

Crystalline: Photovoltaic cells made from a slice of single-crystal silicon or polycrystalline silicon.

CSP (concentrating solar power): is focused sunlight. CSP plants generate electric power by using mirrors to concentrate (focus) the sun's energy and convert it into high-temperature heat (or steam). That heat is then channeled through a conventional generator. The plants consist of two parts: one that collects solar energy and converts it to heat, and another that converts the heat energy to electricity. Within the United States, over 350 MW of CSP capacity exists and these plants have been operating reliably for more than 15 years.

Customer Load: The amount of power your site uses. Load may be expressed in kilowatts (capacity) or kilowatt-hours (energy). A site's peak kilowatts generally refers to when electric demand requirements are highest.

Demand: The level at which electricity is delivered to end-users at a given point in time. Electric demand in measured in kilowatts.

Direct Charges: Those charges directly attributable to a contract or job. They do not include operational expenses, such as overhead, G&A and taxes.

Direct Current (DC): The flow of electricity that flows continuously in one direction. Frequency - The number of cycles through which an alternating current moves in each second. Standard electric utility frequency in the United States is 60 cycles per second, or 60 Hertz (Hz).

Due Diligence: to a potential acquirer, due diligence means "making sure you get what you think you are paying for." This means doing your homework or examination on the offer or situation.

DUNS: Data Universal Numbering System.

Earnings/Share: The after-tax profit of a company divided by the issued and outstanding number of shares.

EBIDTA: Earnings Before Interest, Depreciation, Taxes, and Amortization.

Economic Development Office: In USA, A county governement office, with the responsibility for increasing county revenue from manufacturing, tourism, retirement and related business operations. There exists 3,140 such offices in the 50 USA states.

EE: Energy Efficiency: Improving efficiency by cutting energy use, improving quality, reducing downtime, and reducing waste streams.

EFD = Energy Finance District is a special district created by local government to raise funds to finance the installation of renewable energy systems and permanent energy-efficiency improvements.

Funding for the improvements would be raised from public and/or private sources and loaned to the property owner. The loan would be repaid over a term of 15 to 20 years as an addition to ad valorem tax bill. If the property were to be sold, the new owner would assume repayment of the loan. Participation is voluntary.

Electric Grid: the electricity transmission and distribution system that links power plants to customers through high-power transmission line service.

A power transmission network is referred to as a "grid". Multiple redundant lines between points on the network are provided so that power can be routed from any power plant to any load center, through a variety of routes, based on the economics of the transmission path and the cost of power.

Electromagnetic Environmental Effects (EEE): a broad term meant to include all electromagnetic interference or disturbance – unintentional or intentional, including, but not limited to EMI, EMP, HEMP, HPM, jamming,

Electromagnetic Pulse (EMP): an electromagnetic radiation from an upper atmospheric nuclear explosion that couples into all manner of cables and metallic objects. To test for compliance a field strength is developed at 50 kV/m having a rise time of 5 nanoseconds and a pulse duration of 150 nanoseconds.

Energy - The ability to do work; different forms of energy can be converted to other forms, but the total amount of energy remains the same.

Energy Audit: A survey that shows how much energy used in a home, which helps find ways to use less energy.

EIA: Energy Information Administration: The U.S. EIA collects, analyzes, and disseminates independent and impartial energy information to promote sound policy making, efficient markets, and public under-standing of energy and its interaction with the economy and the environment.

Energy Information Administration
1000 Independence Ave, SW
Washington, DC 20585

National Energy Information Center
(general energy information)
(202) 586-8800
infoctr@eia.gov

Technical Information
(202) 586-8959
wmaster@eia.gov

Fixed Tilt Array: A photovoltaic array set in at a fixed angle with re-spect to horizontal.

Geothermal Energy: energy derived from the warmer (or colder) temperature below the surface of the earth than the inside of the infrastructure being heated (or cooled).

GigaWatt = 1 GW = 1,000,000,000 Watts = 10^9 watts.

Global Warming: See greenhouse effect.

Greenhouse Effect: The carbon pollutants dumped in the air act like a carbon dioxide layer to trap the gases like a greenhouse, thereby warming the earth.

Grid: see electric grid

Grid-Connected System: A solar electric or photovoltaic (PV) system in which the PV array acts like a central generating plant, supplying power to the grid.

Gross Profit: Profit developed from sales in which only the direct charges are applied. This excludes operational expenses (overhead and G&A) and taxes.

HEMP: A High altitude Electromagnetic Pulse generated from an upper atmospheric nuclear explosion. In military terminology, HEMP results from is a nuclear warhead detonated hundreds of kilometers above the Earth's surface. Effects of a HEMP device depend on a large number of factors, including the altitude of the detonation, energy yield, gamma-ray output, interactions with the earth's magnetic field, and electromagnetic shielding and protection of targets.

HEMP is usually described in terms of three components defined by the International Electrotechnical Commission, called E1, E2, and E3:

The **E1** pulse is the very fast component of nuclear EMP. It is a brief but intense electromagnetic field that can quickly induce very high voltages in electrical conductors. The **E1** component causes most of its damage by causing electrical breakdown voltages to be exceeded. **E1** is the component that can destroy computers and communications equipment and it changes too quickly for ordinary lightning protectors to provide effective protection.

The **E2** component is generated by scattered gamma rays and inelastic gammas produced by weapon neutrons. This E2 component is an "intermediate time" pulse that lasts from about 1 microsecond to 1 second after the beginning of the electromagnetic pulse. The E2 component of the pulse has many similarities to the EMP produced by lightning, although the electromagnetic pulse induced by a nearby lightning strike may be considerably larger than the E2 component of a nuclear EMP. Because of the similarities to lightning-caused pulses and the widespread use of lightning protection technology, the E2 pulse is generally considered to be the easiest to protect against.

The **E3** component is very different from the other two major compo-nents of nuclear EMP. The E3 is a slow pulse, lasting tens to hundreds of seconds, that is caused by the nuclear detonation heaving the Earth's magnetic field out of the way, followed by the restoration of the mag-netic field to its natural place. The E3 component has similarities to a geomagnetic storm caused by a very severe solar flare. Like a geo-magnetic storm, E3 can produce geomagnetically induced currents in long electrical conductors, which can then damage components such as power line transformers.

Because of the similarity between solar-induced geomagnetic storms and nuclear E3, it has become common to refer to solar-induced geomagnetic storms as "solar EMP." At ground level, however, "solar EMP" is not known to produce an E1 or E2 component.

Hertz: The unit of electromagnetic frequency that is equal to one cycle per second.

Hydroelectric: Electric power generated by turbines driven from the fall, passage or head of water.

Insolation: The solar power density incident on a surface of stated area and orientation. It is commonly expressed as average irradiance in watts per square meter (W/m^2) or kilowatt-hours per square meter per day (kWh/ (m^2·day)) (or hours/day). In the case of photovoltaics it is commonly measured as kWh/(kWy) (kilowatt hours per year per kilowatt peak rating)

Interconnection: The linkage of transmission lines between two util-ities, or between a utility and an end-user, enabling power to be moved in either direction.

Inverters: Electrical devices used to convert low DC voltage from solar-PV cells or panels to higher AC voltages for direct use in homes and non-residential buildings.

IPO: Initial Public Offering: The first time that a private company has gone public by selling its registered securities.

Irradiance: The direct, diffuse, and reflected solar radiation that strikes a surface. Usually expressed in kilowatts per square meter.

Isolate: An acronym for **Sol**ar bright days, Site **Lat**itude, and electric utility rates. Isolate scores applied to each US state will give a first, quick-look, rough measure of the viability of a proposed or existing solar rooftop installation performance, yet independent of the installation specifics. There are three main parts contributing to overall solar system installation performance:

- (1) Location = site latitude, % bright solar days, and electric utility rates

- (2) Solar technology, mounting and roof configuration
  Financial, prices, costs, subsidies, break even, P&L, cash flow and ROI.

- Isolate score is a rough measure of the state site location in producing affordable solar energy. Its viability score is defined as:

$$\textbf{Isolate} = \textbf{Sol} \times e \times \cos(\textbf{lat}) \times N \qquad (3)$$

where, N is a normalizing/scaling constant and cosine of the latitude, rather than latitude, per SE, is used because that is the way the physics of the math model works.

ITC = Investment Tax Credit: The Fed Gov. offers an ITC to companies and homeowners who install Renewable Energy devices to increase affordability by effectively lower the price.

kilowatt (kW): 1,000 watts. A unit of measure of the amount of electricity needed to operate given equipment. For example, a one kW system is enough power to illuminate 10 light bulbs at 100 watts each. (volts x amps = watts). Or, a one kW system, if operating at full capacity for 5 hours will produce (or use) 5 kWh of electricity.

kWh = kilowatt hour, an energy term = 1 kW of electric power for one hour or X kW for 1/X hours, or any combination of power and time, yielding 1 kWh.

Maximum Power Point (MPP): The point on the current-voltage (I-V) curve of a module under illumination, where the product of current and voltage is maximum. For a typical silicon cell, this is at about 0.45 volts.

MegaWatt = 1 MW = 1,000,000 Watts = 10^6 watts.

Meter: A device that measures levels and volumes of customer's electricity use.

Million = 1,000,000 = 10^6

Moore's Law: A law of electronics technology made famous in an article in Electronics magazine in 1965 by Gordon L. Moore, who would later become the co-founder of Intel. Moore's law states that the number of transistors on an integrated circuit will double in less than 24 months, with corresponding decreases in the cost of electronics technology.

Mounting Equipment: Equipment/apparatus used to fasten solar (PV) modules to the roof.

Multicrystalline: A semiconductor (photovoltaic) material composed of variously oriented, small, individual crystals. Sometimes referred to as polycrystalline or semicrystalline.

NAIC: North American Industry Classification System.

National Electrical Code (NEC): Contains guidelines for all types of electrical installations. The 1984 and later editions of the NEC contain Article 690, "Solar Photovoltaic Systems" which should be followed when installing a PV system.

Net Metering: "Net Metering" is the metering of electricity consumed from the electric utility grid and conversely, exported to the grid (the meter runs backward when excess solar electricity is fed back) by a home or business (office building)

One-Axis Tracking: A system capable of rotating about one axis used to track the sun's daily path in the sky.

Orientation: Placement with respect to the cardinal directions, North, South, East, West. Azimuth is the measure of orientation from north.

Peak Load - The highest electrical demand within a particular period of time.

Peak Sun Hours: The equivalent number of hours per day when solar irradiance averages 1,000 w/m^2. For example, six peak sun hours means that the energy received during total daylight hours equals the energy that would have been received had the irradiance for six hours been 1,000 w/m^2.

Photovoltaic Cell or Module or Panel: (PV) - A device that produces an electric reaction to light, thereby producing electricity.

Photovoltaic (PV) Array: An interconnected system of PV modules that function as a single electricity-producing unit. The modules are assembled as a discrete structure, with common support or mounting. In smaller systems, an array can consist of a single module.

Photovoltaic (PV) Conversion Efficiency: The ratio of the electric power produced by a photovoltaic device to the power of the sunlight incident on the device.

Polycrystalline Silicon A material used to make photovoltaic cells, which consist of many crystals unlike single-crystal silicon.

Power Factor of an AC electric power system is the ratio of the real power flowing to the load to the apparent power (assuming voltage and current are in phase), and is a number between 0 and 1, expressed as a percentage. When PF < 1, the electric utility must send more apparent power, thus, charging more than consumed. Hence, correct the power factor to appx. 100% to save money.

165

Rainwater Harvesting: Collecting and storing rain from structure roofs or other collectors for reuse other than processed for drinking.

Renewable Energy: Energy derived from that which will never run out, such as from wind, sun, rain, rivers, waterways, and heat from the earth, trees and vegetation.

Revenue Streams: From product manufacturers and/ or service companies, on-going regular or periodic (e.g: monthly) sources and amounts of sales revenues.
Examples:

- From carbon credits, such in "Cap and Trade" legislation
- From making business deals with utility companies, clients, & banks
- Continuing local, national international seminars
- Introducing new Trade publication(s) with paid ads.
- From/for municipality clients
- Maintenance and post expiration guarantee services
- Special EMP-mitigation hardware protects clients from catastrophic
    vulnerability of national electric grid.

ROI = Return on Investment: Net moneys received above an invested sum. ROI = (total present value – original investment)/ (original investment).

Sector Identification: level of renewable energy installation ranging from an individual home or small commercial-office building to municipalities.

SGEMP = System Generated EMP: When gamma and x-rays from a high altitude detonation encounter a satellite in space they excite and release electrons as they penetrate the interior of the system. This phenomena is referred to as system generated electromagnetic pulse (SGEMP) because the accelerated electrons create electromagnetic transients. Systems must be configured with special cables, aperture protection, grounding, and insulating materials in order to survive these transients.

SGEMP impacts space system electronics in three ways. First, x-rays arriving at the spacecraft skin cause an accumulation of electrons there. The electron charge, which is not uniformly distributed on the skin, causes current to flow on the outside of the system. These currents can penetrate into the interior through various apertures, as well as into and through the solar cell power transmission system. Secondly, x-rays can also penetrate the skin to produce electrons on the interior walls of the various compartments. The resulting interior electron currents generate cavity electromagnetic fields that induce voltages on the associated electronics which produce spurious currents that can cause upset or burnout of these systems. Finally, x-rays can produce electrons that find their way directly into signal and power cables to cause extraneous cable currents. These currents are also propagated through the satellite wiring harness.

Shell Company: a public corporation which has discontinued its business but holds cash in its treasury hoping to be acquired by an attractive, growing private company in order to go public.

SID, See Sudden Ionospheric Disturbance

Smart Grid: Electric grid stake-holders have identified the following characteristics or performance features of a smart grid:

- Self-healing from power disturbance events (except EMP)
- Enabling active participation by consumers in demand response
- Operating resiliently against physical and cyber-attack (Questionable)
- Providing power quality for 21st century needs (Questionable)
- Accommodating all generation and storage options
- Enabling new products, services, and markets
- Optimizing assets and operating efficiently (Questionable)

Solar: Energy from the sun shining on collectors whose elevated temperature is used to heat water for pools, hot-water heaters, and other applications..

Solar-Earth Seasons: Since the earth is tilted on an 23.5° axis as it rotates around the sun in a year, four seasons are developed which greatly impact the efficiencies of solar installations. The seasons begin:

- Vernal Equinox, 21 March = day and night of equal length. Sunrise = location latitude

- Summer Solstice, 21 June = longest daylight of the year. Sunrise = location latitude + 23.5°

- Autumnal Equinox, 21 Sept. = day and night of equal length Sunrise = location latitude

- Winter Solstice, 21 December = shortest daylight of the year. Sunrise = location latitude - 23.5°

Solar Energy: Heat and light radiated from the sun.

Solar Flares: the eruption from active sunspots which produce excess radiation during the solar 11-year sun-spot cycle. Among other disturbances, they impact ionospheric propagation which can disrupt communications, especially in the HF-VHF (3 MHz – 300 MHz) spectrum.

A solar flare is a sudden brightening observed over the Sun's surface or the solar limb, which is interpreted as a large energy release They are often followed by a colossal coronal mass ejection also known as a Corona Mass Ejections (CME). The flare ejects clouds of electrons, ions, and atoms through the corona into space. These clouds typically reach Earth a day or two after the event.

Solar flares affect all layers of the solar atmosphere (photosphere, chromosphere, and corona), when the medium plasma is heated to tens of millions of kelvins and electrons, protons, and heavier ions are accelerated to near the speed of light. They produce radiation across the electromagnetic spectrum at all wavelengths, from radio waves to gamma rays, although most of the energy exists at frequencies outside the visual range. For this reason the majority of the flares are not visible to the naked eye and must be observed with special instruments.

The frequency of occurrence of solar flares varies, from several per day when the Sun is particularly "active" to less than one every week when the Sun is "quiet", following the 11-year solar cycle. Large flares are less frequent than smaller ones.

Strong solar flares can happen at any time, but are more common during the peak half of the 11-year sunspot cycle, can cause wide regional blackouts to the electric grid. Since they are a very low frequency phenomena, long high-voltage, electric power lines act as pick-up antennas resulting in the possible burnout of transformers at substations since transformers are not designed to handle the very low frequency (nearly DC) currents induced in the windings.

Solar, First Generation: Crystalline Silicon solar represents the most popular with about 85% of the solar installations as of 2010. It is the most expensive and about 20% efficient in converting sunlight into electricity. It is the oldest of the technologies and, typically, has a 25 year guarantee and a 40+ year life expectancy.

Solar Panel: Devices that collect energy from the sun (solar energy). This is usually solar photovoltaic (PV) modules that use solar cells to convert light from the sun into electricity, or solar thermal (heat) collectors that use the sun's energy to heat water or another fluid such as oil or antifreeze.

Solar-P: Energy from the sun shining on cells converted to DC electricity by photovoltaic action. Through inverters, this is then converted to AC electricity for driving electrical loads.

Solar Resource: The amount of solar insulation a site receives, usually measured in $kWh/m^2/day$, which is equivalent to the number of peak sun hours.

Solar, Second Generation, Thin-Film solar is made from amorphous silicon (the least favorable) or popular cadmium tellurium (Cd Te) or CIGS in thickness less than a human hair, but mostly formed into rigid, glass-covered panels. Efficiencies are about 11%, but reader beware as this is not the correct measure of better performance (see below).

Solar, Third Generation, Inks and Dyes: mostly in experimental stages and a few to several years away from practical, competitive installations.

Solar Thermal: The process of concentrating sunlight to create high temperatures that are needed to heat fluids, like water (solar hot water) or to vaporize fluid to drive a turbine for electric power generation.

Solar Water Heating: Using the sun's rays to heat an absorber material which transfers the increased temperature to buried pipes carrying water.

Every solar water-heating system features a solar collector that faces the sun to absorb the sun's heat energy. This collector can either heat water directly or heat a "working fluid" that's then used to heat the water. In active solar water-heating systems, a pumping mechanism moves heated water through the building. In passive solar water-heating systems, the water moves by natural convection. In almost all cases, solar water-heating systems work in tandem with conventional gas or electric water-heating systems; the conventional systems operate as needed (night or overcast days) to ensure a reliable supply of heated water.

Stand-Alone System: An autonomous or hybrid photovoltaic system not connected to a grid. May or may not have storage, but most stand-alone systems require batteries or some other form of storage.

Storage: Storage refers to saving surplus electricity produced by a photo-voltaic (PV) system. Generally, batteries are used as storage devices.

String: A number of photovoltaic modules or panels interconnected electrically in series to produce the operating voltage required by the load.

Sudden ionospheric disturbance (SID): an abnormally high ionization/plasma density in the D region of the ionosphere caused by a solar flare. The SID results in a sudden increase in radio-wave absorption that is most severe in the upper medium frequency (MF) and lower high frequency (HF) ranges, and as a result often interrupts or interferes with telecommunications systems.

When a solar flare occurs on the sun, a blast of intense ultraviolet and x-ray radiation hits the dayside of the Earth after a propagation time of about 8 minutes. This high energy radiation is absorbed by atmospheric particles, raising them to excited states and knocking electrons free in the process of photo-ionization. The low altitude ionospheric layers (D region and E region) immediately increase in density over the entire dayside. The ionospheric disturbance enhances VLF radio propagation. Scientists on the ground can use this enhancement to detect solar flares; by monitoring the signal strength of a distant VLF transmitter, sudden ionospheric disturbances (SIDs) are recorded and indicate when solar flares have taken place.

Short wave radio waves (in the HF range) are absorbed by the increased particles in the low altitude ionosphere causing a complete blackout of radio communications. This is called a short wave fading. These fadeouts last for a few minutes to a few hours and are most severe in the equatorial regions where the Sun is most directly overhead. The ionospheric disturbance enhances long wave (VLF) radio propagation. SIDs are observed and recorded by monitoring the signal strength of a distant VLF transmitter.

Tracking Equipment: Structure that houses PV modules and that can automatically follow the sun across the sky throughout the day to maximize output.

Utility: The interconnection of electricity generation plants through the transmission and distribution lines to customers. The grid also refers to the interconnection of utilities through the electric transmission and distribution systems.

Volt (V): The amount of force required to drive a steady current of one ampere through a resistance of one ohm. Electrical systems of most homes and offices use 120 volts. (volts - watts/amps) (volts = amperes x resistance)

Watt (W): Electric measurement of power at one point in time, as capacity or demand. For example, light bulbs are classified by wattage. (1,000 watts = 1 kilowatt).

Waveguide beyond cutoff: A condition in waveguides (or any other metal tube including air vents), when a frequency from a potential interfering source, f_{EMI}, somewhat below that corresponding to a half wavelength (of the waveguide or tube), propagates with significant, but predictable attenuation.

A first approximation of the attenuation is, Adb = 30 l/d, where l = tube length and d = tube diameter or width. To assure 80 dB shield attenuation, used several places in this book, make l/w greater than 3.

(for details and exact equation, see, *The EMC, Telecom and Computer Encyclopedia Handbook*, third edition, by Don White, emf-emi control, 1999.)

Wind Turbine: is a rotating machine which converts the kinetic energy in wind into mechanical energy. This is then converted to electricity via a turbine or generator.

Appendix B

Ten Tips to Reduce Cost of Electricity up to 60%

Some solar experts say, "For every dollar you put into energy conservation, you can save $3-$5 in the cost of producing your own power."

While going solar is not going to make one wealthy, it is good as an anti-pollutant for the environment, will increase the value of your home, makes us less susceptible to the increasing costs of energy, and not hold USA hostage to foreign oil. As you use our solar calculator, we ask that you think about all the benefits, and not just the bottom line cost and break even. So, it is important in the decision process to identify what you are *trying to optimize* in the first place.

Ten Tips to Reduce Your Electric Bill

First, see remarks on Insulation and "The 3-minute rule," (Part 1).
Then, come the 10 tips (Part 2).

Part 1- Insulation

The home insulation status is the largest single factor con-attributing to your electric bill. Fig. 1 shows the five principal areas that must be examined for adequate insulation.

If the house is already built, the options for saving are less. Start by focusing on the heat loss in the attic. Attic fans and extra insulation blown-in or flat fiberglass insulation on the attic floor to attain an R-30± rating is the first step. For basements used in the Northern latitudes in USA, use a 1-2" solid foam between the concrete blocks

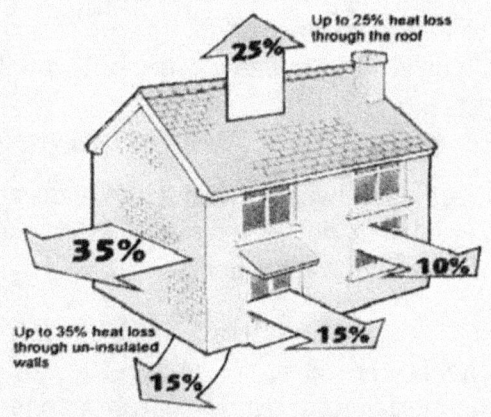

Fig. 1 House showing the five areas of primary heat and cooling loss.

and the wall-board or knotty pine or other decorative finish to an R-12 rating. Some new insulating paints are available with an R-4 rating.

Leakage at outside door-to frame regions rank second in loss. Change or modify their insulation. Replace any single-pane windows with at least double pane types. Since this latter expense is usually, at least schedule their replacement over a specified time.

Figure 1 also shows other areas which can then be addressed for insulation upgrading.

The Three-Minute Rule

With one exception, this conservation rule requires some discipline and involves no additives. The rule states.

"If you are the only person in a room, watching TV, using electric lights, operating an electric device, and the like, turn off all items if you will be gone for more than three minutes. Exceptions may be a computer under certain conditions (put in a sleep mode instead) or something which may require a reset time or re-operative conditions. If only for a "few minutes", do not turn off and on, as there is some life-expectancy stress in surge, unless a surge suppressor is used in the power line.

The benefits from *the three-minutes rule* may result in your electric bill savings from 8% to 22% (maybe more), if you do nothing else. Thus, in a typical $150/month electric bill, you may save $240±/year.

Part 2 – 10 Tips for Electric Power Conservation

Electrical power is measured in watts and kW; 1 kilowatt = 1,000 watts. Cumulative power over time is measured in units of energy or kW-hours = kiloWatts x hours.

1. Hot-Water Heater (HWH).

The HWH consumes about 10% of your electric load. Reduce its thermostat to about 110° from a typical 125° setting to save wasted money. Search around the house for old blankets and towels to wrap around the heater tank and secure for improved insulation. Or, buy a preformed blanket made specifically for this purpose.

Keep the HWH circuit breaker "off" until 30± minutes before you need hot water for the clothes washer, dishwasher or for personal showering.

When you are finished, switch off the breaker.

2. Air Conditioning and Heating Savings

Set the AC to 78° F or somewhat higher in the summer to save electricity. Turn it a bit higher at night or more (say 82°) when away for ½ day or longer. Setting the AC much higher to save more electricity is not advised because of the danger of mold buildup in the Southern coastal states.

When heat pumps are used for winter home heating (even for oil, gas or coal furnaces), set the daytime room temperature to 68°-72°. During the night, upon going to bed, reset the thermostat to 65°. Actually, most new homes in the past 10-15 years have a programmable thermostat. Thus, the different settings can be done to cover a month or a season. Honeywell seems to be the leader in this sector, and they are carried at both Home Depot and Lowes.

3. Washer and Dryer

The Washer–Dryer represents a large electrical load. The washer costs about $0.04 per load for electricity and about $0.12/load for water. For hot water, the cost is $0.16 + about $0.30 for heated water = roughly $0.40/load. EPA says the average family uses 7.6 loads/wk ($13/month). The big saving is to use fewer loads and at cooler (or cold) water temperature. Have you ever tested to see how well clothes get cleaned using cold water? You may be surprised.

The dryer is the big electrical load. The average drying load consumes roughly about $0.34 in electricity For the average EPA loads of 392/year, the dryer cost $133/year to run. Here, consider some air drying like on wood racks or close lines if no home deed restrictions.

In summary the washer-dryer combination costs about $0.75/load or $295/yr. Cutting down the number of loads, using cold water washing, and drying on racks will save much, if not most, of this cost.

4. Dishwasher

Most of the savings here is not to run the dishwasher unless it is fully loaded with dirty dishes. Small loads, or dishes needed again soon, can be hand washed and placed in a drainage rack to dry. The cost will also be reduced if the hot-water heater was set to 110° F instead of 125° as mentioned above.

5. Pool and Irrigation Pump Motors

Sprinkler or lawn irrigation systems use 1-1.5 horse-power motors and pool pumps also use about 1-1.5 HP motors. A pool heater, when applicable, should use roof solar or butane gas heat) since an electrical heater is very expensive to operate and could double your electric bill (depending on your latitude).

One horsepower in a motor or pump rating corresponds to 746 watts. If you made the mistake of buying a heat pump for your pool, sell it and replace it with a rooftop solar heating or gas heater with the proceeds of the electric heater. You will save big time and use the pool longer as most users quickly learn how expensive it is to electrically-heat a pool.

Many owners will run a pool pump for 4-6 hours in the summer (to combat algae), and 2-3 hours in the winter if in the South (zero in the North as the pool was winterized. If these numbers are reduced to 3 hours in the summer and one in the winter, approximately 60 kWh will be saved in the summer and 45 kWh in the winter. This approximates 630 hours/year, or 950 kWh for a 1.5 HP motor. At an average cost of 13 cents/kWh, the owner in the South will save about $122/year. The Savings in the Northern latitudes is far less.

6. Incandescent Light Bulbs

Traditional incandescent light-bulb filaments are subjected to inrush stress current thereby limiting their life expectancy. Incandescent bulbs from 25 watts to 100 watts can be replaced by the same size fluorescent bulbs which produce comparable light intensity for 25% of the electricity consumption. Fluorescent bulbs also have 10 times longer life expectancy. This is a very compelling reason to replace many if not all incandescent bulbs in the house.

For example, suppose the family room and other areas collectively use eight incandescent bulbs totaling 600 watts for six hours a day. The cost is 600 watts x 6 hrs x $0.13/kW x .001 (kW/watt) = $0.47/day. If you used fluorescent bulbs, the cost savings would be 0.75 x $0.47 = $0.35/day = $10.53/month = $126/year.

The average cost of an incandescent bulb approximates $0.35 and the cost for its fluorescent mate is roughly $1.50 or four times more. However, the life expectancy of the incandescent bulb is less than 1,000 hours, while that of the fluorescent exceeds 5,000 hours. Thus, their replacement costs per lifetimes are somewhat better for fluorescence. Meanwhile the user enjoys one fourth the electricity consumption. (Ed. Note: I have observed about eight fluorescent bulb burn outs in a four period. Thus the life expectancy is questioned).

7. High Duty-Cycle Appliances

Duty cycle means the percentage of the time in use. If the house has two refrigerators, such as one in the garage or for outside patio use, consider disconnecting the outside one as it is expensive to use in the summer (it is also much hotter outside than in the house). Set the kitchen refrigerator to warmer temperatures if it is at or near the coldest setting, which is not needed.

The heating and A/C system duty cycle is also high in the summer and winter, but low in the spring and fall. It was addressed in Subsection 2 above.

8. Low Duty-Cycle Appliances

If the oven and burners are electrical, not much can be done to reduce the electric bill, other than to turn them off when finished using. These are the big kitchen loads, so don't heat up until you are ready to use. The microwave oven shuts off when a selected period for cooking is over. So when the MW can do the lighter cooking, us it in favor of the oven and burners. Remember, don't turn on the MW oven with no load placed therein as damage may result or keystroke life shortened.

Turn off the coffee maker after use or put it on a timer. Some people let it run for hours; then, turn it off.

A ceiling fan takes about 100 watts or about 2-5% of an A/C load. Run one or more it at low speed, if needed. This can reduce some of the A/C duty cycle. This also keeps hot food from cooling on the counters.

9. Den and Office

Most computers today are laptops (15-45 watts; desk-tops consume 60-250 watts) and consume about four times the power of a desktop computer. The LCD screen in a laptop consumes most of the power, typically about 30 watts.

Many laptop computer owners add a 19-23 inch LCD external screen as it is easier to read than smaller print. The power consumption is correspondingly more (roughly 50 watts). Modern computers have internal "sleep" modes when not in use for a while (they power down to 3-5 watts), which time is internally settable. Thus, when leaving a computer for a half hour or longer, place it in the sleep mode. Turn it off during the night when not in use.

Supporting printers, copiers, scanners and the like should be turned off when not in use. Some standby modes waste power.

10. Outside Lighting

Nearly all outdoor lighting should be put on timers which cost less than $10 each. With the new fluorescent light bulbs, where appropriate (flame shaped decorative incandescent bulbs do not yet have a fluorescent counterpart), the cost to operate outside lighting be-comes 25% of the older incandescent bulbs.

Finally, solar outdoor lights tend to become a waste of money as the batteries have to be replaced each year at $2 apiece or $4 per lamp. Low-voltage AC lighting is more cost effective since the bulbs last much longer than the battery life of solar outdoor lighting.

The Bottom Line:

Of course, if you are already doing many of the above electricity-saving conservation steps, your options to save will be considerably less. Otherwise, when all the dollar savings above are added up for the typical $150/month electric bill, they approximate $65-$80/month or $850±/year. This is equal to 47±% savings. Note: this is also roughly equal to about what you would save if you invested in a solar rooftop electric system saving equal to 50% of your electric bill. And, except for the insulation cost mentioned above, it costs you nothing except, possibly, a change in yourself discipline attitude about conservation.

Appendix C

Third Generation, Solar Paint and BIVP
(for technical readers)

The following is about solar paint and represents information abstracted from articles appearing in Cal Finder website and *Solar Products* and *Renewable Energy World* magazines and others displayed on the Internet in the Spring, 2012.

Solar paint is literally a paint containing a layer of dye-sensitized cells that could be coated onto steel and eventually turn millions of steel buildings worldwide into innate solar collectors. From UK universities and a steel manufacturer, *Corus Group*, solar paint is one step closer to the day of mass production.

The technology has been developed and, starting October 2010, a laboratory in North Wales began work to develop and produce the new solar paint. The final product will have four different layers: an undercoat, the layer of dye-sensitized cells, electrolytes, - titanium dioxide, a white pigment, and a protective film as the final layer. All layers combine to form a white paste that can be rolled onto steel sheets during manufacturing.

Scientists and researchers are basically mimicking natural photo-synthetic processes. Light illuminates hits the dye-sensitized cells and excites molecules in the cells which, in their excited state, release electrons into the titanium dioxide layer. There, the electrons are collected and produce electrical current. Then, the electrons are attracted back to the dye by positively charged diodes, regenerating the cells. Another advantage of these cells is that they can absorb light across a wider range of the spectrum; good news for areas, such as the UK, with a relatively high percentage of cloudy days.

There has been little said about life expectancy and warranties as the product is not yet manufactured and installed to start the life cycle per-formance measurements. If present 20-year guarantees are required of new products, it may be many years before installations will appear in significant numbers.

The following is a summary of an article appearing in Oct 2010 on the CalFinder website:

Third Generation Solar Photovoltaic Cells

The **third generation solar photovoltaic cells**, still relatively new, mostly involves cutting edge nanotechnology - the manipulation of molecules and atoms at an extremely small scale. The conventional crystalline solar collectors are considered 'first generation' and the thin film PV (amorphous silicon) is 'second generation'[1] The highest efficiency achieved in laboratory test conditions is 24%. [2] . "The best silicon PV modules now (2004) available commercially have an efficiency of 17%, and it is expected in about 10 years' time (2014), module efficiencies will have risen to over 20%[3] . While progress is slow but steady, there are 'third generation' solar cell researchers attempting to raise the bar of efficiency. These innovative approaches to solar energy harvesting may hold the potential of achieving great leaps in solar module efficiency. Some of theses technologies are in the testing phase but most are still under exhaustive research.

Paint-on semiconductors: quantum dots

Over half of the sun's energy reaching the earth is invisible to us because it is delivered through infrared wave lengths. Current solar collectors only work under direct visible sunlight, "and if we don't tap into [infrared light] in our solar cells then we throw away more than half of the potential sun's energy that we could be using[4] ." The great advantage of this technology is that it captures infrared energy during the day - even when the sky is cloudy. The basic building block of this technology is called the quantum dot. Quantum dots are nanoparticles consisting of extremely small (a billionth of a meter) collections of atoms of semiconducting material[5] . It would take a chain of 4 million quantum dots to span the length of a penny. These are the first solar cells that are able to harness the sun's infrared rays. The plastic material uses nanotechnology to improve the cell's efficiency. This breakthrough has led theorists to predict that plastic solar cells could one day become five times more efficient than current solar technology[6] ."

"Now nanotechnologists are breaking away from large, perfect, crystalline semiconductors. Instead they are building physically flexible solar fabrics. These photovoltaics are printed like newspapers, spinning seamlessly from roll to roll."[7]

Ted Sargent speaks about infrared solar technology
Credit: You Tube

Liquid silicon technology

This process involves the conversion of silicon from a solid into a liquid in order to increase the utility if the semi-conductor. The technology takes tiny silicon particles (in powder form) called "nano-crystals" and chemically liquefies it into ink. At this point the substance can be used in a regular inkjet printer and solar cells can be printed on any surface. "This is a completely radical way to think about how a semiconductor material can be deployed[8]." This appears to be one of the most promising of the 'third generation' technologies. "This is a dramatic break though because it means that almost any surface can become a solar cell. We can paint silicon materials onto roofing and bring down the cost of having solar installations on you roof by a factor of 10[9]."

Dye-sensitive solar paint

Dye-sensitive solar paint is a solar cell technology that is currently in the development stage. This photovoltaic paint "will be based on dye-sensitive solar cells" not silicon, making the technology relatively inexpensive. The difference between conventional **solar panels** and dye-sensitized solar paint is that "instead of absorbing sunlight using silicon…[it] uses dye molecules attached to the particles of the titanium dioxide pigment [found] in paint. Once light hits the dye, the electrons are 'activated' and they jump from the dye into a layer of electrolyte.[10]

Conrad Burke develops liquid silicon, Credit: You Tube

The energy created from the activated electrons is then transferred to a collecting circuit before the electrons return to the dye, completing their cycle. There are some researchers are experimenting with fruit based

179

dyes. After separating the dye from the heavy particles and sugar molecules, the dye is placed on a "conductive glass-coated with a film of titanium dioxide (the common material found in paint)."[11] This bonds the dye to the glass.

The installation process would be simple and can be applied directly to industrial steel. "The solar cells are built up in several layers. Firstly, a barrier of normal paint is laid directly on the steel, then the electrolyte and dye layers, and finally a clear protective film to guard against the elements."[12]

This technology could eventually be used with the steel sheets that are traditionally used to cover large buildings. The greatest advantage of this technology will be its low cost and its ability to be applied as a paste. Even though the solar paint is less efficient than the panels, its competitive advantage is reflected in its reduced cost.

Implication for energy policy

Third generation PV solar cells could become the most reliable source of clean energy if research efforts are adequately funded. "Eventually if research on third generation PV proves successful, it could lead to PV cells made, for example, from extremely thin stacked plastic sheets, converting solar energy to electricity with very high efficiency at very low cost[13]." Some suggest that the technology might "double or triple the 15%-20% range currently targeted[14]" technology may one day provide energy to the 2 billion people on this planet that live without electricity.

Footnotes:
1. Boyle; pg.78, 82
2. Boyle, Jeffery. Renewable Energy:Oxford university Press, 2004; pg 68
3. Boyle; pgs 68
5. Bole; pg. 82
6. Sargent, Ted: Interview, http://www.youtube.com/watch?v=kLSARSw2JUQ
7. Sargent, Ted. The Dance of Molecules: How nanotechnology is changing our lives. Thunder Mountain Press. New York; 2006
8. Burke, Interview: http://www.youtube.com/watch?v=UPpvPAriAOY
9. Fred Krupp, President of Environmental Defense

Fund: http://www.youtube.com/watch?v=UPpvPAriAOY

10. Marshall, Michael. Solar-power paint lets you generate as you decorate. New Scientist; Tech, March 2008: http://www.newscientist.com/article/dn13424

11. http://solar.calfinder.com/blog/news...d-solar-oh-my/

12. Marshall, http://www.newscientist.com/article/dn13424

13. Boyle; pg. 82

14. Green, M.A., Third generation photovoltaics: Ultra-high conversion efficiency at low cost. Progress in Photovoltaics: Research and Applications, 2001. 9(2): p. 123-135. doi:10.1002/pip.360

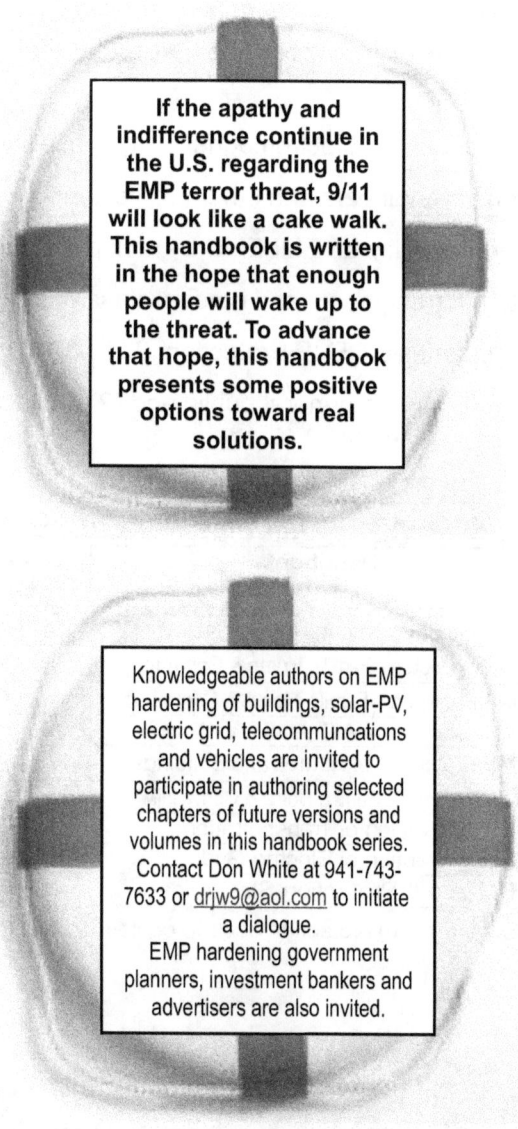

If the apathy and indifference continue in the U.S. regarding the EMP terror threat, 9/11 will look like a cake walk. This handbook is written in the hope that enough people will wake up to the threat. To advance that hope, this handbook presents some positive options toward real solutions.

Knowledgeable authors on EMP hardening of buildings, solar-PV, electric grid, telecommuncations and vehicles are invited to participate in authoring selected chapters of future versions and volumes in this handbook series. Contact Don White at 941-743-7633 or drjw9@aol.com to initiate a dialogue.
EMP hardening government planners, investment bankers and advertisers are also invited.

The Handbook Series on EMP Protection

This book you are now reading is (the third edition of) the first volume of a planned series of five volumes on EMP and Solar Threats and protection of the targeted infrastructure. As seen in the table below, some of the books are written for the general public and some for the engineering and related professional world.

This will be the legacy of the author, Donald R. J White, and other younger researchers and writers engaged by him. This monumental project will be the third time in his life that the author has undertaken such an ambitious project:

- Five-volume series on Electromagnetic Compatibility, late 1970s

    written for engineers and other professionals

- Twelve-volume series on EMI/EMC by multiple authors, late 1980s

    written for engineers and other professionals

- Five-volume series on EMP Protection, 2013-2015

    written (1) for the general public and (2) other professionals

2012-2015 Handbook Planning Production Schedule

Vol. Nbr.	Ver. Nbr.	Handbook Title	Read for	Publication Date
1	1.0	Nuclear EMP Threats - What Next?	Lay & Pro	July, 2012
1	2.0	EMP Protection - Family, Home & Community	Lay & Pro	Nov., 2013
1	3.0	EMP Protection – Family, Home & Community	Lay & Pro	May, 2013
NA		EMP Contrast – A Tale of Two Towns	Novel	May, 2013
2	1.0	EMP Protection of Homes and Buildings	Arch & Engr	Sep, 2013
1	4.0*	EMP Protection – Family, Home & Community*	Lay & Pro	Jan., 2014
3	1.0	EMP Protection of Transportation	Lay & Pro	Apr, 2014
4	1.0	EMP Protection of Electric Grid	Engineers	Oct., 2014
5	1.0	HPM Use & Protection Strategies	Military	Jan., 2015

* First EMP (and solar rooftop) protection book to contain advertising

*Planned Advertising in the Next Edition of this book
to be published in January 2014 or possibly earlier*

The idea to accept advertising in a bound book was created and developed over 13 years ago when Don White, the author, wrote *the EMC, Telecom and Computer Encyclopedia Handbook*, Third Edition, published by *emf-emi control*. This 800-page compendium, had 121 advertisers, and thousands of copies were sold worldwide.

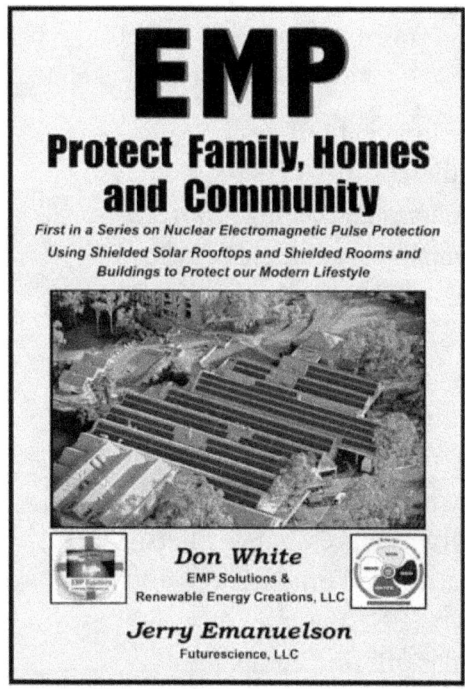

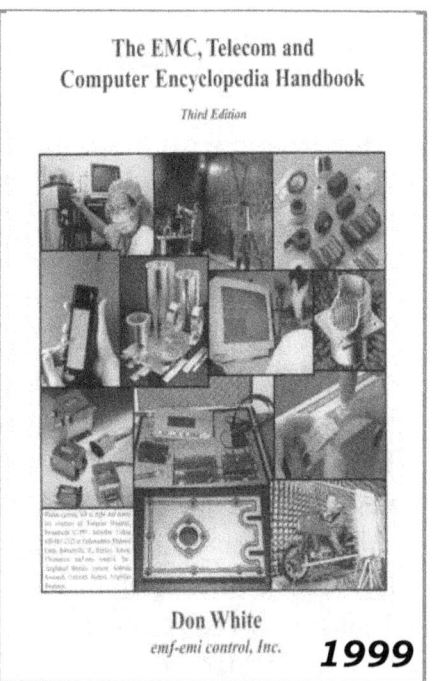

That handbook, displayed on the right, was so successful, we are now planning to repeat this feat in the related EMP and associated Solar-PV energy fields. Thus, 25 pages of planned advertising (about 11% of book) will be inserted in the new 2014 book to be published in January 2014. The cover is displayed at the left for the current 3rd edition.

Promotion and Circulation
Based on past experience, expected softbound book circulation numbers range between 5,000 and 10,000 copies. Fifty advertisers will be

distributing an expected 2,000 book copies alone to their most favored clients.

In the digital world, the EMP book copies will be sold by Amazon's Kindle (their position requires exclusivity). Digital copies are forecast to outsell softbound copies by 2:1 or 3:1. Thus, when added to the above distribution of softbound copies, total circulation is estimated to between 15,000 and 30,000 copies. While it is possible to go below the low side, the chances of exceeding the highside are considerably greater

Buyers of these books include:

- 3,140 U,S. County Economic Development Offices
- Corresponding Chambers of Commerce
- Relevant Federal Government Agencies and their Libraries
- 165 leading US electric power companies and their libraries
- Advertisers who are known to each sell hundreds via give-
  aways to their clientele at symposia, conferences & expos
- Ads in Google, Facebook, Craiglist and other Internet
- Renewable Energy Trade Organizations and their Libraries
- EMP Survivalists including all Internet websites
- EMC Community manufacturers who provide the EMP
  protection shielding, bonding, surge suppression & filters (ITEM)
- Hundreds of manufacturers of solar-related products (ENF)
- Thousands of solar system installers (ENF)
- Dozens of solar training and seminar offering companies
- Distributors and larger selected book stores (8% of the 10,000)
- 5% of largest 104,000 high schools and universities in US
- Selected office building, shopping mall, hotels/motel owners
- Selected foreign solar manufacturers and trade organizations. (ENF)

Additional Benefits of Book Advertising vs. Magazine and Trade Journal.

- Book ads are conspicuous and unique
- Book ads have between 6 and 20 times shelf life
- Book ads have a far greater book content backup

than a few magazine articles per issue.

• Book ads occupy 10-15% of the entire content vs. 50-60% for magazines. Thus, book ads are about 4 times more probable of being read.

Remember, the shelf life of this book is at least two years. Compare that to monthly **magazine** ads having a shelf life of 2-3 months. This is a 10:1 benefit to the advertisers. So divide our ad rates by ten when comparing other magazine options to advertise. You will quickly see the huge yield-cost benefits in going the book ad route vs. magazines.

Early Bird Rate Card as of Spring – Summer 2013 for Fall 2013 Book Ad Placements

Ad Size	Ad Dimensions English Units	Ad Dimensions Metric System	Back of Book	Center of Book	Inside Covers
Full Page	5" wide x 8" high	127mm x 203 mm	$950	$1,250	$2,250
1/2 page, horiz	5" wide x 4" high	127mm x 102 mm	600	750	NA
1/2 page, vert	2-7/16 w x 8" h	62mm x 203mm	600	750	NA
1/4 page, horiz	5" w x 2" h	127mm x 51mm	360	450	NA
1/4 page, vert	2-7/16 w x 4" h	62mm x 102mm	360	450	NA
1/8 page	2-7/16 w x 2" h	62mm x 51mm	235	295	NA

Terms 40% with order, 60% upon publication: First come, first served until all space gone
Early ads may appear in beta, prepublication copy. Ad Closing date 20 Dec 2010

The following table lists the early bird ad rates for our new EMP book. A 40% payment gives the advertiser a larger best choice of location. Since the early bird rates end on 30 September, the higher seasonal rates start on 1 October and end on 20 December 2013. Publication date is 30 January 2014.

Ad Rate Sizes

The next three pages show the ad sizes in the book:

Perfect-Bound (softbound) Book
Measures 6" x 9"

This is a Full Page Ad

It measures 5" x 8"
(127 mm x 203 mm)

Early Bird Rate Card as of Spring – Summer 2013 for Fall 2013 Book Ad Placements

Ad Size	Ad Dimensions English Units	Ad Dimensions Metric System	Back of Book	Center of Book	Inside Covers
Full Page	5" wide x 8" high	127mm x 203 mm	$950	$1,250	$2,250
1/2 page, horiz	5" wide x 4" high	127mm x 102 mm	600	750	NA
1/2 page, vert	2-7/16 w x 8" h	62mm x 203mm	600	750	NA
1/4 page, horiz	5" w x 2" h	127mm x 51mm	360	450	NA
1/4 page, vert	2-7/16 w x 4" h	62mm x 102mm	360	450	NA
1/8 page	2-7/16 w x 2" h	62mm x 51mm	235	295	NA

Terms 40% with order, 60% upon publication: First come, first served until all space gone
Early ads may appear in beta, prepublication copy. Ad Closing date 20 Dec 2010
4th Edition to be Published on or before 30 Jan 2014

This is a ½ page
vertical ad
It measures 2-7/16" x 8"
(62 mm x 203 mm)

This is a ¼ page
vertical ad
It measures 2-7/16" x 4"
(62mm x 102 mm)

Early Bird Rate Card as of Spring – Summer 2013 for Fall 2013 Book Ad Placements

Ad Size	Ad Dimensions English Units	Ad Dimensions Metric System	Back of Book	Center of Book	Inside Covers
Full Page	5" wide x 8" high	127mm x 203 mm	$950	$1,250	$2,250
1/2 page, horiz	5" wide x 4" high	127mm x 102 mm	600	750	NA
1/2 page, vert	2-7/16 w x 8" h	62mm x 203mm	600	750	NA
1/4 page, horiz	5" w x 2" h	127mm x 51mm	360	450	NA
1/4 page, vert	2-7/16 w x 4" h	62mm x 102mm	360	450	NA
1/8 page	2-7/16 w x 2" h	62mm x 51mm	235	295	NA

Terms 40% with order, 60% upon publication: First come, first served until all space gone
Early ads may appear in beta, prepublication copy. Ad Closing date 20 Dec 2010

This is a 1/8 page ad
It measures 2" x 2-7/16"
(51 mm x 62 mm)

This is a 1/2 page horizontal ad
It measures 5" w x 4" h
(127 mm x 102 mm)

This is a ¼ page horizontal ad
It measures 5" w x 2" h
(127 mm x 102 mm)

About Placing an early bird ad order now

Early bird advertisers will benefit from the lower ad rates and best location selection inside the book, by placing their ad(s) before 15 July 2013. and making the 40% down payment before 30 August 2013. All partial payment receipts are deposited into an escrow fund used exclusively for next edition advertising. Fill out the order form below (later to appear on our website, www.emp-safeguard.com) and submit to EMP Solutions and Renewable Energy Creations, LLC.

Company name:_____

Contact person: _____ Title:_____

City:_____ State:_____ Mail Code:_____

Country:_____. Tel. #:_____ Fax #:_____

e-mail:_____@_____. Website:_____

Select your ad location and size:

☐ Back, full pg: $950   ☐ Back ½ pg: horiz $600   ☐ Back: ½ pg vert $600

☐ Back 1/4 pg horiz $360   ☐ Back ¼ pg vert $360   ☐ Back 1/8 pg $235

☐ Cntr, full pg: $1,250   ☐ Cntr ½ pg: horiz $750   ☐ Cntr: ½ pg vert $750

☐ Cntr ¼ pg horiz $450   ☐ Cntr ¼ pg vert $450   ☐ Cntr 1/8 pg $235

☐ Inside front cover: $2,250   ☐ Inside back cover: $1,500

Your total advertising commitment for _____ ads = $_____
    40% binding commitment = $_____ due before 15 Aug. 2013

_____   _____, 2013
        Authorized Signature   Date

(1)- Tear out this page and mail or copy and e-mail as attachment to
    drjw9@aol.com

(2)- We confirm by fax or other mail

(3)- You receive confirmation within 48 hours

Book Advertising is unique and finds its way into all manner of outlets. Long shelf life, too

Request Your Handbook Evaluation

Please take a few minutes to score this handbook as your results will be used to update the next edition, expected about every year. Yes, you will be benefiting next readers and the entire community based on your feedback. Unless you wish to remain anonymous, we will credit each new edition with those improvements based on your suggestions. Spare no pain, good, bad or indifferent. And you will receive a free signed next edition if your feedback is referenced. Thank you.

Name:_____. E-mail:_____. Tel #:_____

(1)- Score handbook with 0-10 score. Then, score each box with your evaluation as follows:

10 = Excellent, 8 = Very Good, 6 = Good, 4 = Fair, 2 = Marginal, 0 = Poor

Handbook Topic Evaluation Scores

Topic ID	Score	Topic ID	Score
Overall handbook		Overall quality of images	
Overview idea for each chapter		Lay ability to understand	
Quality of discussions		Impact anything you might do?	
Sufficiency of detail		Quantity/quality of Index	
Large number of images		Usefulness of Appendices	

(2)- Best and Worst of this handbook:

Which topics or subjects are the most important to you? _____

_____

Which topics or subjects are of least interest to you? _____

_____

Where the handbook was of greatest value:_____

How to make it better yet:_____

Where the handbook was of least value:_____

Where the handbook got it wrong or needs correction:_____

_____

Other recommended improvements for the the next edition:

_____

(3)- Since EMP hardened, store provision-replenishment vehicles are vital to keeping food, water, medications and fuel available to survivors, transportation is the subject of Vol. 3 of this EMP series. Comments please:_____

_____